GOD'S PLAN OF CREATION

A SERIES books 3

The Story:

The Theology of Cosmology

Book 3 The Story:

The Theology of Cosmology

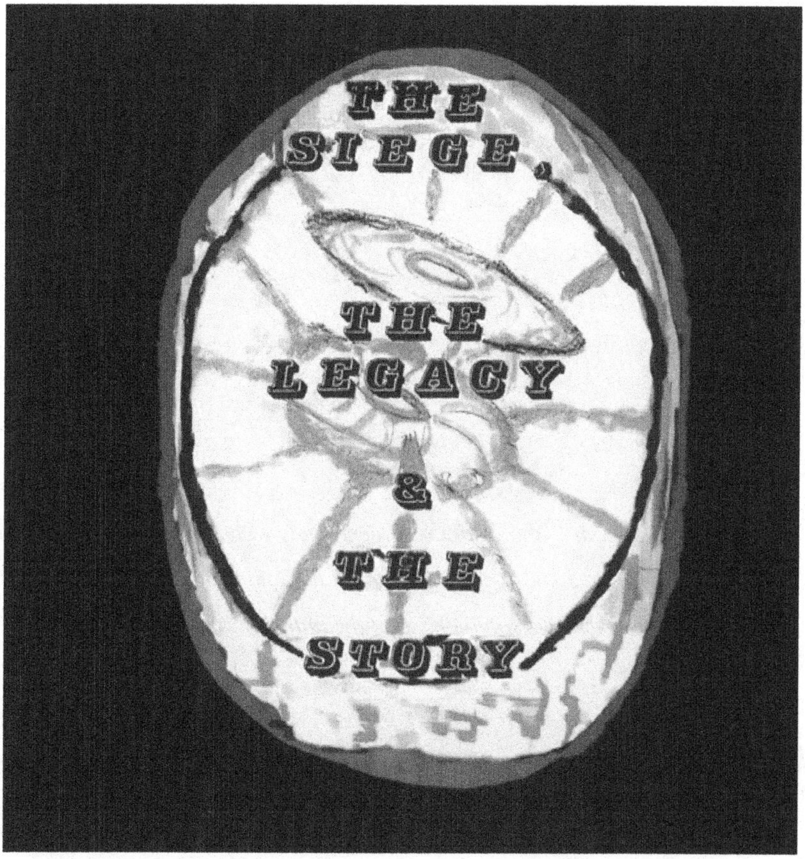

Book I The Story of the Siege

Book II The Story of the Legacy

Book III The Story: Theology of Cosmology

Published 04/16/2018

ISBN: 978-1-7337917-4-8 (sc)color
ISBN: 978-1-7337917-6-2 (sc)b&w
ISBN: 978-1-7337917-5-5 (e)

Library of Congress Control Number: 20179322

Table of Contents

A SERIES books 3 i

Table of Contents v

A Short List Of References: x

How This Book Was Written xii

The Siege xii

The Legacy xii

The Story xiii

Why This Book Was Written xiii

The Vision xiii

The Dreams xiv

The Voice xv

Recognition xviii

Introduction 1

The Story A Timeline 3

Looking From Two Different Perspective 5

The Big Question 6

Three Important Definitions. 7

An Ancient Hindu Story. 8

We Are All Sons And Daughters Of God. 8

What Is Consciousness? 9

The Birth Of Our Soul 12

The Law Of Cause And Effect, Karma 14

Choices Of The Soul 15

Time Is Relative To Our Perspective. 18

Understanding God's Plan Of Creation 22

God Divided Creation Into Three Divisions: 23
1. Paradise: 23
2. The Central Heaven 23
3. The Heaven of heavens 24
The Four Cornerstones 28
The One God That Became Three 29
The Eternal Isle Of Paradise 31
Maltese Cross of Paradise 35
The Isle of Paradise 37
The Central Heaven 38
Stepping Down Of The Eternal Triune 43
The Physical Elemental Cosmos 45
Coming Forth Of The Seven Super Universes 46
Scientific Look At Our Universe 47
The Three Great Heavens 48
Expansion Of Creation 49
Deity Presence Gravity 50
Parallel Dimensions 51
Progressive Dimensions 51
Progressive Dimensions 52
The Involution From Spirit To Matter 53
1. CONSCIOUSNESS 54
2. MATTER 55
Two Spiritual Dimensions 56
God Is Spirit, And That Is Different Than Matter. 57
Top-Down Of Creation 58

Overview Of Creation....................59
The One God....................59
Paradise....................59
The Light & Darkness....................60
The Central Heaven....................60
Two Great Belts Of Dark Bodies....................60
The Heaven Of Heavens....................61
One Of Seven Heavens Within The Heaven Of Heavens....................61
Local Universes (A Galaxies)....................62
Informational Data of Jesus, Kingdom....................62
Numerical Organization And Population....................64
The Speed Of Light....................65
Approximate Size Of Our Galaxy....................65
The Local System....................66
Orgazinatol chart of Christ's Kingdom....................68
The Beginning of the Time Line....................69
When The Thrones Were Cast Down....................70
The 7th. Heaven Our Super Universe....................71
With the Primary Nebulae and the Mother Mass....................73
The Secondary Nebular Stage....................74
The Mother Mass Of Nebadon....................75
The Great Prince Michael Arrives On Salvington....................76
Two Phases Of Creation Were Simultaneously Being Created....................77
Organizational Chart of Lucifers Domain....................78
The First Step Of The Second Nebular Stage....................79
The end of the great Andronover Nebula....................80

The Birth Of Our Solar System 81
Development Of Urantia Our World 82
Early Earth Developing..84
Urantia Was Registered 85
The Life Dawning Era...86
The Flood Period Genesis 1: -20-23 87
Primitive Mankind Appears On The Earth 90
Lucifer Rebellion: The War In Heaven 95
Gabriel Fights War In Heaven 96
Time Line Of Adam & Eve's Descendants 99
Cain's Descendants 100
A Break In The Time Line...105
The Spiritual Development Of Mankind 108
Preparations For The Bestowal Of A Spirit Born Son Of God 110
The Mindset of Early Civilization..112
Understanding Early Mankind's Mindset 113
Mindset Of Today...115
The Birth, Childhood, And Youth Of Jesus The Young Boy. 117
Jesus Becoming Aware Of His Destiny...133
The Great Disillusionment At The Temple...133
Jesus Experience During The Passover 135
Jesus Left Behind In Jerusalem 136
Jesus' 4th Days In The Temple 137
Jesus' In The Temple 137
Jesus Slowly Removes Himself From Family,....................................150
Jesus Takes One Year As A Caravan Leader 154

Jesus' Ministry 157

The Two Gospels Of Good News 159

Two World Orders 160

Where We Are At Now 161

World Order For The Brotherhood Mankind 162

The sovereignty of the ONE GOD 162

Sovereignty Of The Individuals Will 162

Sovereignty Of Collective Humanity 163

A layout for World Order 165

The Global Order for the Brotherhood of Mankind (GOBM) S 165

The Religious Freedom Branch (RFB) 165

Should be one or two 165

The Legislative Law Branch (LLB) 166

The Peace Keeping Branch (PKB) 166

Conclusion 167

A Short List Of References:

foremost the BIBLE:

King James Version
The New Oxford anointed Bible
The Jerusalem Bible
New Marked Reference Bible
The New American Bible
The Catholic Holy Bible
World Bible
Catechism of the Catholic Church
The Triumphant Church
Divisions of Dying and Rebirth
The book of Witnesses

Ancient, and other Books

The Dead Sea Scholes
The Aquarian Gospel of Jesus Christ
The Gospel According to Thomas
St Martin de Porres
The Nag Hammadi Library (and its many scrolls)
Forbidden Mysteries of Enoch
Pistis Sophia
Destined for the Throne
The Kabbalah
The Universal Meaning of the Kabbalah
The Koran
The Dhammapada
Bhagavad-Gita (as it is)
The Upanishad (as it is)
Hindu Scriptures
The Secret of the Golden Flower
Upanishads, Gita and Bible
The Foundations of Tibetan Mysticism
The Tao-The-King sayings of Lao Tze

Cosmic Consciousness
Altered States of Consciousness
The Nature of Personal Reality
Psycho-Cybernetics
Chariot of the Gods?
The Sayings of Confucius
Zohar
The Greek Philosophers
Jonathan Livingston Seagull
Illusions
Mysteries and Legends of the North American Indians

SPECIAL note for the following
My books will run parallel to the vast information contained in the following and to whom I owe a lot of thanks.

The Urantia Book from many books. Pamphlets and videos.
https://www.urantia.org/

Edgar Cayce A.R.E. Foundation from many books. Pamphlets and videos.
https://www.edgarcayce.org/

Al Miner and Lama Sing foundation from many books. Pamphlets and videos.
https://lamasing.org/

These last three references have a vast amount of information from which we may draw from that will enhance our spiritual growth.

God's origination and the timeline followers along with The Urantia Book.

The spiritual growth and development follow the advice laid down by the Edgar Cacey ARE foundation, where I studied for almost two years.

I also had the fortune to know Al Miner and Lama Sing, through several readings and a continued study of the information from their origination.

All their information was and is to assist humanity in raising their consciousness of their relationship with God and prepare the way for the full fulfillment of the Promise.

How This Book Was Written

The Sieg*e*

The Siege is a discussion of many religious traditions and dives deeply into the war that took place in heaven and the spiritual war on earth and humanity. The method used takes issues in a question-and-answer format in the manner of puzzle pieces scattered about a tabletop. Suggestive answers are given, tying the pieces together with possible answers. The various subjects should provoke thoughts.

The Legacy

The Legacy is written in a straightforward but suggestive manner, attempting to tie all the discussions of The Siege together in a meaningful pattern. It discusses how we are created in God's image and how we can learn to communicate with our loving Father in heaven. It discusses the process of going to and fro—the only thing Jesus chewed out the Pharisees for not doing or teaching others who wanted to learn. The Legacy will also provoke thoughts on the complexity and cohesiveness and simplicity and completeness of creation.

The Story

The Story starts off with the big question: Is the monolithic force just an energy force or was it with consciousness? The Story is presented as a timeline of creation. It briefly covers the history of creation from before the beginning of time and space. It attempts to lay out how the One God bestowed Himself in the beginning. It was the great bestowal—the mystery of mysteries that brought forth everything needed for all His creation. It briefly covers the creation of Paradise, the Central Heaven, and the Heaven of Heavens. It covers the development of our universe (our galaxy) and how humanity developed the religions of our ancestors. It ends with a timeline of the life of Jesus of Nazareth. As with The Siege and The Legacy, the Story will also lead to thoughts. Perhaps The Siege, the Legacy, and The Story will inspire readers to seek out further answers for themselves.

Why This Book Was Written

There were three main reasons. One was a vision, the second was a series of dreams, and the third was a voice.

The Vision

I received a book to read. I didn't like to read because I had a muscle in my eye that would start twitching after I had read about one page. I would then see a double image every time it twitched. I became a very methodical reader—one word at a time and very slowly.

I took the book just to appease the lady who gave it to me. When I was bored one night, I began to read the book. It was a subject that did not agree with my childhood's religious teachings. I started to throw it against the wall in defiance, and as it passed my ear, I saw a vision. That vision was just as clear as it could be. It was a side view of the universe. Instantly, I knew what it meant: "Don't you dare throw that book away. Keep reading, and the universe will open up to you." Being the good military man I was at that time, I gave a hand salute and said, "Yes, sir." I slid down in my seat with the book in my hand.

My eye never twitched again (to this day), and I read more in the next three months than I had read in my entire lifetime (about three small books). I was about thirty years old then.

The Dreams

Interpreting my dreams has been easy for me since the first time I tried it. I would wake up in the middle of the night, write them down, and interpret them. I had been working with my dreams for some time, and one night, I had a very important one. It has led me on a life-long siege into the mysteries of life.

In my dream, a man came to me and said, "The children are lost in the woods." We left the bar and went into the woods. Everyone was running around and calling the children. Instead of calling, I whistled loudly. When I whistled, the children came out of the woods. Everyone was happy, saying, "When you called, the children found their way out of the wilderness."

I interpreted that dream as saying if I spoke on the mysteries of God's creation, which is what I was studying at the time, God's children would find their way out of the darkness and into the light. They were lost because they did not seem to understand the mysteries.

I thought *this is great. Interpreting a dream is a lot of fun.* The next night, I had the same dream, and I knew it was getting serious. Having the same dream is like using an exclamation point at the end of a sentence instead of a period. The very next night, I had the same dream again. This time, I was in China. I knew it was getting serious. That night, as I said my prayers, I said, "Lord, I realize there is nothing beyond your power. If there is something you want to tell me, then come right out and say it because I could make a mistake in the interpretation of my dreams."

The Voice

The next morning, I woke up and dozed off again. I then became consciously aware of vibration all around me—from head to toe. If I had been moving physically with the vibrations, I would have been bouncing off the walls, floor, and ceiling at the same time. I heard a voice flowing with the vibrations. It spoke with complete authority, saying, "Tell the story of the siege. Tell the story of the legacy. Tell the story." Then it was quiet and still. I was tingling all over from the words flowing through me as I wrote them down. I didn't know what a siege or a legacy was. They were not part of my vocabulary at that time. I hurried to a dictionary to look up the words. It took almost an hour to find those words because I couldn't even spell them.

I combined the dreams with the words I had heard and felt and came to the conclusion that I had been told what to do or asked what to do. It was a tall order for a simple man. I said a quick prayer: "Lord, I asked you to tell me if there was something you wanted me to do, but I didn't expect this. Are you sure you've got the right man?" After a short pause and no response, I said, "If that's what you want, then I'm not going to say no. I only hope You know what You're doing because I don't."

God seems to delight in taking the ignorant and making them smart, and taking the dumb and making them wise. I must have been one of His greatest delights. I have since completed high school, earned an AA degree and a BA degree, and read more books than I can remember. I can lay no claim to any great knowledge or wisdom. I lay no claim to being a saintly person. The Lord knows I have as many shortcomings as anyone else. I do, however, lay claim to two things: I love the Lord my God with all my heart and soul, and I trust Him implicitly. So, trusting and loving God—and with the greatest love and respect for every man, woman, and child of every race, color, and creed—I have prepared this book. I hope it accomplishes the task of telling the story of ***"The Siege," "The Legacy," and***

"The Story."

Allow me to share with you a vision I had after I started writing this book. I was in prayer and meditation when an angelic force appeared in front of me. He handed me a book. I took the book in both hands; it was a big book with henges; I put it in my(big) mouth, and swallowed. I ate the book. That was the end of the vision. I often thought about it, but I never told anyone because I didn't know what the book was about. I didn't ask for the title; I just ate it. I often wondered why the angel never gave me another book. I imagined that the angel was in heaven telling everyone, "Don't give that dummy another book. All he will do is eat it."

If you are eating a book while in the spirit has a couple of meanings, first, it means that you consumed the book. Second, it means that you ingested the contents. We know that, when we ingest food in the physical, we digest it and then eliminate it. Ingesting a book in the spirit means that you will digest its contents in the mind, heart, and soul—and then it will come out in the form of writing, speaking, and/or actions.

After a few years, I forgot about the vision. At about the twenty-year point, I had been blindly writing down everything I had been given. I did not know what the book was going to be about until I wrote the part of this book titled "The Divine Destiny of Humanity." I was thinking about the overall flow of the book when the vision came to me again. This time, I understood what the book I had eaten was discussing. It was amazing. I knew the name of the book. I could not believe my gracious Father had given this simple man such a wonderful book. I cannot swear to this because I did not see the title, but I have a gut feeling that the name of the book I was given to consume was *God's great Plan of Creation.* As you read, decide for yourself.

At the twenty-year point, I thought, *At last, I'm done.* Instead, I was sent back to the drawing board. I continued to write, correct, and rewrite for another twenty years. At the forty-year point, the purpose of the book finally became clear to me. I think it is a stepping-stone for when the world will move into the light. When the world moves into the light, there will be so much to learn and grasp that it will boggle the minds of many people. I hope this book will soften the blow of the heaviness of all there will be to learn. If anyone can begin to understand this book, it may make it easier to take the giant steps into the wisdom and knowledge of God. I hope this book will serve as a small stepping-stone for humanity, my brothers and sisters.

Recognition

So many people have aided me in the preparation of this book. It would take another book to list them all. I will acknowledge a few of them. Arlene Shafer gave me the book *Many Mansions,* which led me to the vision. Thank you.

I picked up many pearls of wisdom in Virginia Beach, Virginia, where I was led by my dreams to meet Joan and Ralph Sides. We were a part of a Search for God group, a Bible study group, and dream interpretations sessions. After the experience described in "Why This Book Was Written," I started writing this book. That was a joke in itself because my spelling, punctuation, and grammar were so bad that people who couldn't spell laughed at me. I wrote persistently, knowing that it would be finished one day.

In another Search for God group in the Jacksonville Beach area of Florida, I met Elda McMann and her sister Ray. Ray took a lot of my chicken-scratching and typed it into a readable form. Joann Weaver took another part and typed it again and again. Mrs. Denmark did the same with another section. The effort of these ladies put the material into a readable form that was ready for a rewrite.

I started typing the book on my computer, and that was another joke. If I rush, I can type minus ten words per minute. After struggling with "The Siege," I realized I might not live long enough to see the book finished. I turned to the source that never failed and asked for help. Lo and behold, a friend named Eugene Nagasawa and his son David could type, and they agreed to type the rest for me.

The book was now ready to go to school. After a few phone calls, I was led to a lady who made so many fine corrections that she must have worn out several pencils. After the corrections had been made, it was ready for a final reading. After a few more corrections, the book was ready to be considered for publication.

After a few years and several rejection slips, I figured there must be something wrong. I went back to the source that never fails, placed the book on our altar, and said, "I think the book is finished. I hope you are happy with it." That night, in my dream, a policeman (a symbol for higher authority) came to me and said, "No, it is not." I said, "Yes, sir," and I went back to the drawing board. He was right. I was not finished. As I went back to work, I received more information. It has been the most wonderful and mind-boggling experience that this old country boy has ever had. Thank you, Lord.

At this time, I would also like to add a very special thanks to Al Miner and Lama Sing for their warm words of encouragement.

What started out as vibrating words, and the appearance of a joke has become a reality after nearly fifty years.

To all of these wonderful people and others, I extend my heartfelt warmth and appreciation. Most of all, I want to thank God for allowing me to write this book. Without God's help, guidance, assistance, and love, this book would not be.

Introduction

The siege is an inquiry of man's continual effort to gain the legacy that has been willed to us by God and passed on to us by many great souls. God created us in His image. All of which were discussed in Book I & Book II.

This third book ia the story is a timeline of creation that combines theology with cosmology.

Special Notes To The Reader

The siege is an inquiry of man's continual effort to gain the legacy that has been willed to us by God and passed on to us by many great souls. God created us in His image. All of which were discussed in Part I & Part II.

The third part is a timeline that tells the story. This book is full color and is the conclusion of part 1 & 2. It starts before creation began and ends when Jesus took off his work apron and said, "My time has come."

The story is a timeline of creation that combines theology and cosmology trying to show their relationship in the process of creation. It is not intended that all of the information is to be taken as absolute nor definitive. It is hoped that it will enlighten each reader enough that they open their minds and assimilate the information and help them formulate their own opinions, their own hypothesis

For the student of the Bible, I have used the King James Version. It has the reputation of being the best translation from the original books chosen for the Bible. Also, there have been many significant changes to other God's plan of creation.

Gods Plan of Creation
A SERIES

Book - 3

The Story A Timeline

Theology of Cosmology

Part of Michelangelo's creation of the heavens, Sistine Chapel ceiling.

WRITTEN FOR ALL OF GOD'S CHILDREN WHO WANT TO KNOW

Looking From Two Different Perspective

SCIENCE	RELIGIONS
HAS BEEN LOOKING FOR THE ONE (**MONOLITHIC**) FORCE OR POWER OR GROUP, THAT ACTS AS ONE, FROM WHICH ALL CREATION CAME AND IN WHICH ALL CRERATION EXIST.	HAS BEEN LOOKING FOR THE **ONE GOD** WHO IS THE INEFFABLE OMNI-OMNI (THE ALL PERFECT PRINCIPAL OF ALL PERFECT PRINCIPLES) ESSENCE FROM WHOM ALL CREATION CAME, AND IN WHOM ALL CREATION EXIST.

This is a study of the theology of cosmology. Trying to answer WHY, along with HOW, that monolithic force of creation came into being. It is the hopes that the thoughts expressed herein will provoke more thought and result in a deeper understanding of these two great endeavors that are trying to find our relationship to and in that magnificent plan of creation.

The Big Question

If we go back in time and space, all the way back to where there was **no** time **nor** space, we would be in the presence of what scientist would call a MONOLITHIC force or power.

A Monolithic force or power is one power or a group of powers that acts as ONE, from which all creation came and in which all creation exists.

Science would say it is the **only** power. Spirit-minded and religious people would say, "Yes, it is one power; we call it the ONE GOD." By calling it God that makes it a deity, which in turn means it is alive with consciousness.

Therein is the **big question**: Is that one force or power *alive with consciousness*, **or** is *it just an energy force without consciousness?*

The only way to investigate it is to go there using our consciousness, which is that quality that science says the One did or does not have. Our consciousness is that mental process we reason with and ask questions then search for answers. It is that aspect of our being that is aware of being aware of being conscious and aware of our surroundings.

So let us take that journey. Back to where there was only that one force or power, some of us call The One God.

First, let us take a scientific look. Science says energy cannot be destroyed; it can only change. That monolithic power would have to be an unmeasurable amount of energy or an unknown power that did, or could, produce an almost unmeasurable amount of energy far greater than the big bang explosion. That unknown force would have been brewing like a stew from before eternity. If there is such a thing. It was there moving on its power until something happened that caused it to change, an accident. That change would create a chain reaction to take place within that nebulas body of energy. That would be the cause of that which is called creation. It began to unfold purely by accident and without a conscious decision.

Now let us take a look from a spiritual point of view. That monolithic force would have been in existence for eternity like our scientist suggest, except it would have been with an innate consciousness, content within itself. At some point, it became aware that it was alone and decided to create a companion, one like unto itself. Causing a manifestation of creative force all designed to bring about an organized creation wherein new life could be born and grow into companions with the One.

Three Important Definitions.

Let us look at three definitions before we get too deep in the subject matter.

1. **PRIMORDIAL FORCE**: The primary force from which all creation came and in which all creation exists.

2, MONOLITHIC FORCE: The one force, or group of forces or powers, that acts like one from which all creation came and in which all creation exists.

3. ONE GOD: The DIVINE Omni-Omni essence (the all perfect principal of all perfect principals) from whom all creation came and in whom all creation exists.

An Ancient Hindu Story.

The One God looked around and saw that He was alone. At first, He was frightened, then He reasoned, what is there to be afraid of? There is only ME.

We Are All Sons And Daughters Of God.

The ONE GOD was with an innate ability to know and recognize itself as being consciously aware of having consciousness, and conscious of his surroundings. In other words, He was aware of being alive. The One God has always been aware of being self-contained within himself. Now He desired companionship. One God thought, planned, and decreed His plans of creation wherein new life could be born and grow into companions with Him. book 3 is the story of that plan of spirit and matter into which He bestowed (gave) Himself.

What Is Consciousness?

Simply stated, it is the state of being aware of one's self and our surroundings. All of us can make the statement "I am conscious, I have consciousness." Each of us has a lot of differences, yet we have many similarities. We may be of many national origins, faith, color with many, many differences, yet we can all say, "I am part of humanity." The total of our consciousness, all of us, is what makes us what we are. We can say the same about all peoples of all nations. Combined the consciousness of all the people of the world would equal to the consciousness of the world. We can follow that process all the way to that Monolithic Force we call God. If everything came from that one force, then it must have had consciousness and would have been alive. If it did not have consciousness, then we would not. We do, it did.

That brings up many other questions. How did that consciousness get from that One to here? Why? How are we related to that One? Why? How did we as an individual get here? Why are we here now in this small insignificant distraught world? The list of questions could go to another page or so, but I'm sure you get the idea.

Science, with a sweep of the hand, would say the One God did not exist. We have no need of the hypothesis of God; we can figure it out ourselves. (Sounds a little arrogant.) They would say there is no such thing as consciousness; it is but a figment of our imagination. It is an electrochemical reaction of the fleshly thing we call the brain. Too often, we are fooled into thinking that it exists. Most scientist would be among the first to say they are alive and aware of being aware of their self and their soundings. Hum! That is ironic, isn't it?

They do have free will just as we all have free will. We can all believe what we want. After all, it's just a belief. A belief is something we think is true. We spirit-minded and religious people know the truth by the lessons we have experienced, both positive and negative; they become truth to us as lessons learned. Science knows something to be true by the data they have collected, which allows them to create hypotheses. Science is always scrutinizing their findings. They seek to prove or disprove their hypotheses or findings. They are trying to fine-tune the truth down to a gnat behind. Too bad religions do not borrow a little of that fine-tuning in their search for the One God. No one likes a belief forced on them, scientific nor religious. That philosophy is, "It's my way, or you will be damned." After all, we are free-willed creatures, and we all have the right to choose for our self what we want to believe. It's our choice.

We are all just wimps trying to figure out this universe and how we are part of it. A joke, of course, but not too far from the truth. Because like the wimp, we seem infinitely small and insignificant. Each of us is a fragment, like a cell from the body of the One God. We were created in the beginning as a spiritual child when the One God called forth the light, a spiritual Light, which was ALL the heavenly host being called into existence. That monolithic force, (with consciousness) bestowed Himself. That bestowal was the first requirement, according to the great plan of creation. All of the creation of spirit and matter was laid out before Him and all of us, His spiritual children. All of us were given freedom of choice, our free will. That act was the birth of our soul (and the mind the soul's companion). Together they have a complete recollection of all the adventures and experiences since that moment. That spirit within has a complete awareness of the total essence of the One God. Some souls chose to return immediately and are called angels. Most of us chose to venture out into an experience in this new creation. The Universe is a big backyard for all of His children to play, live, and grow experientially as an infant child grows into adulthood. If you read with an open mind, you may discover how it and how you are a part of that great plan.

The Birth Of Our Soul

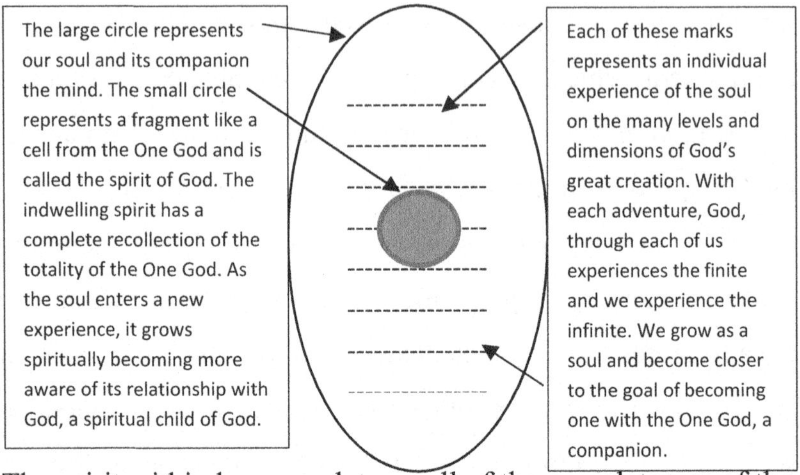

1. The spirit within has complete recall of the completeness of the One God. It is of, or from, the One God. Via that spirit, we are all equal, equally a part of God and equally loved by God.
2. The soul has complete awareness of **all** its many experiences, but not of the totality of the spirit. As an entity in each of those experiences, we have a recollection of what we learn in that experience, which becomes a part of the totality of the soul.
3. With each experience, positive or negative, (some would label good or bad), but in truth, we learn from both, and that is good in God's eye.
4. With each experience, we learn and grow spiritually. Slowly but surely, each of us progresses closer and closer to becoming like unto the Father and a companion with the One God. The spiritual center will grow until the totality of the soul is spiritualized.

5. God has given us all eternity to accomplish that task and return to Oneness with Him.
6. Science has been looking for that primordial monolithic force from which all creation came. If they find it, would they see the life forces in it? Probably not: This is ironic because they are among the first to acknowledge that they are alive, aware of their existence, and capable of thinking about their environment and the cosmos. Since they could accept that all creation came from that primordial monolithic force, then the life forces in them would have come from that one monolithic spiritual force or power, we call God. If everything came from nothing, it would return to nothing. Nothing comes from nothing. It would have no purpose. All our learning, growth, and experiences would amount to nothing. That is a concept with no meaning. If there was a creation, there must have been a creator.
7. If it came from the One God, then it will return to God as a companion. God, in His infinite wisdom and understanding, has created a spiritual and elemental universe with many dimensions and levels wherein we, His sons and daughters may experience, learn, and grow. Via our free-will, we choose where we wish to venture. Once we make that choice, we are obliged to finish therein. God entrusts or gives us a portion of Himself, like a seed. Together we grow. God experiences the finite, and we experience the infinite, as we grow into companions. We are all His sons and daughters; therefore, we are all brothers and sisters.

We should act with respect and brotherly love toward one another.

8. Wisdom, knowledge, understanding, intuition, strength, courage, and worship are the seven columns or tools used by the Spirit to teach us. Respect, love compassion, mercy, goodness, kindness, living in right standing with others, and God is, what creation is all about. This is what we should be working on, not self-gratification, self-aggrandizement, self-glory, nor selfishness.

The Law Of Cause And Effect, Karma

Without a control system freewill would run rampant. Freewill is governed by the law of cause and effect. Our actions and deeds directed by our intention will activate that LAW. It is like unto a double-edged sword. On one side, we expect to be punished if we do something considered to be wrong. On the other hand, we expect to be rewarded if we do something considered to be good. If we do those acts with either of those intentions, they will follow us around thought out eternity, because we created them, they are ours. We will be required to meet them and receive their reward or punishment. We may choose to change them by repenting, forgiving, or redirecting them by creating a new cause. That law will be fulfilled one way or another.

This law has been stated as an eye for an eye, a tooth for a tooth, a limb for a limb, a drop of blood for a drop of blood, and a life for a life. It has been paraphrased by the Golden rule: Do unto others as you would have others do unto you. Harm not others with that which harms you. Do not wish for others that which you wish not for yourself.

The motivation we should have is love, compassion, and mercy for all God's children, our brothers, and sisters. We should not be looking for a return or expecting a reward. We should only try to be of service, with respect for others right to their freedom of choice, because that was given to all of us. If our attitude is governed in this manner, there would be no enactment of the law of cause and effect, only rewards which will be granted by our Lord God as He sees fit. His love and compassion for us are far beyond our comprehension. His gifts are so great that our minds cannot conceive their depth nor width.

Choices Of The Soul

Freewill simply means we are given freedom to choose for ourselves when, where, and how we want to experience God's wonderful creation, that big backyard which is all those spiritual levels and all the elemental creation. Once we decided and we applied our will--- it began. At first, it was an act of getting involved in what was happening. We either obeyed the universal laws of creation, working with God, or we venture out on our own.

The more we practiced the art of creating, the more we got involved in that which we created. We choose the place and when. We choose what we wanted to experience and how. We were not restricted to just the spiritual dimensions; we could venture into and out of any dimension, including the elemental universe as well.

It didn't take long, figuratively speaking, before we began to forget from where we came, and that we were spiritual children of God. We were free-willed souls on a long and adventurous journey, enjoying every moment of it.

Somewhere along the journey, we began to realize it was harder to return. We were getting deeply involved in God's creation because of those things we ourselves created were affecting our vision, our thoughts, and were muddying the pits we were getting involved in. However, we were still having a lot of fun until we started to replace our thoughts about others into thoughts about self. But, that did not stop us, we continued to dive head-on into our own desires, creating more and more difficult for ourselves. We were learning how to use and misuse those precious gifts from God. Somewhere along the line, we looked around and realized we were lost; we could not see where we came from. We forgot our way back home. So, we continued to experience as if nothing had happened, until we began to experience the power over others. We experienced anger, jealousy, and hate. We began to experience that we were superior or inferior to others. The negativity of all sorts was flooding our souls. We were really getting lost in the darkness of forgetfulness. Then it happened, we were bogged down in the muddy pits of our own doing. The last act that locked the door to return was spilling the blood of another. We were completely lost. Completely involved in the earth. The seeds were planted.

As we looked around, we could hardly see it was so dark; we could feel it pressing in on us, making it hard to breathe. As we called out for help, we could see the light far in the distance twinkling as if it was saying follow me; I still remember the way back.

At that point, we decide we have had enough. It's time to return---it's time to go back home. That light began to speak to our heart and give us advice on how to get out of the entanglement that we ourselves created. We must return via the route we came. We must set those crossings of God's universal laws aright. We must meet and untangle all those creations that were not just and right between us and our brothers and sisters. We must choose of our own free will to change our thoughts, attitudes, and we must begin to act right toward God and our brothers and sisters.

It has been a long and reckless trip out; it is a long and tedious trip back. We will be directed every step of the way by those of the heavenly host. We will begin to ascend the ladder returning back to the Father. There will be many choices to make. Each step we take makes us wiser children, grown into adulthood, becoming companions with our loving Father, the Only Son, and the Eternal Spirit, who will lead us in the final steps of becoming companions with the One God.

Time Is Relative To Our Perspective.

We view things from our point of view here on our planet. The earth is spinning around on its axes creates our days. The earth is also revolving around the sun, which creates our year. The equatorial bulge and the tilt of the planet as it revolves around the sun create our seasons. What if you could perceive our local system of stars or the movements of our constellation, how would we calculate time then? What if you could perceive the movement of our galaxy or cluster of galaxies or even a supercluster of galaxies, how would you calculate time then? What if you could perceive the movements of the whole universe? Let us take this thought another step. What if you could perceive all things of the elemental universe and all the spiritual dimensions above them, how then would we perceive time? Time then is relative to our perspective.

It is hoped that this paper will give the reader a better understanding of creation, both spiritually and physically. Therefore, we will look at creation from two different perspectives.

First: **Theological**: in that Creation, is far too coordinated and organized to be happenstance. It probably was created by the primordial force a Supreme Being or force, which we call the ONE

GOD, who has established control of all his vast creation. That is too simple an answer for science. If there was a creation, which is obvious, then there must have been a creator. Nothing comes from nothing, yet we say God is Omni-powerful if He wanted to create everything from nothing He could, but did He?

Second: **Cosmological**: Evolving simply means a slow, tedious process of changing. If one looks, it becomes self-evident that everything is evolving, and everyone is evolving or going through that slow, tedious process of changing. We are not the same as we were when we were born; neither is the universe. If we would try to combine these two ideas together, it may become obvious that both are right. Therefore both ideas of creations should be looked at as a magnificent and glorious work of art that is spiritual and physical, and scientific and religious, that was created by a maker, and it is continually going through that slow, tedious process of changing. It is amazing how much our scientific brothers have discovered about creation. They look for "how" while religious, and spiritual sources look for "why." Some people say all that science has discovered serves to prove Gods existence. Science is uncomfortable when looking for why.

Imagine if you will, a vast nebulous body with no specific form, but with consciousness, aware of existing and being alive. As the One conceived something, it would appear from the awesome spiritual energy of the nebulous body. As the One God realized He was alone, He desired companionship. As the One planned out all of the creation and thought the thoughts, the movement began taking shape within the spiritual energy of the nebulous body. He spoke the words, that which he had thought began to appear. They were becoming a reality on that level that has become known as Paradise.

Let us dive into the depth of creation and put together some of the pieces of the puzzle of that great plan. We should always remember we are not all at the same level of consciousness and not under the same conditions. But, we are all equally endowed with the precious gift of free will. Freedom to choose. So, with respect and love for all humanity, let us put on our boots and jump into the spiritual and elemental cosmos.

The next plate will give a quick, simple, display of the complex first steps of the GREAT BESTOWAL of the One God. The breaking down and reconstructing His energies and giving Himself into all of creation.

BEFORE CREATION

There was this one force, one power; one essence. That one force, one power, one essence is what science would call the **primordial force** or monolithic force. Spirit minded people would say it was alive and give it personality, character with a consciousness and awareness of actuality and reality. Spirit minded people call it **the One GOD**. The One GOD is the ineffable omni-omni essence from which all creation came and in whom all creation exists. The One GOD is a multi-dimensional creator who has created a multi-dimensional creation. We are multi-dimensional creatures existing in that multi-dimensional creation trying to understand our multi-dimensional creator. Creation was first conceived in the consciences of the One GOD. God planned out every aspect, every detail of creation. He left nothing to chance, but He did leave a lot to choice. With infinite wisdom and knowledge God planned out all of creation.

THE BEGINNING OF CREATION

Religious people say God's plans were nurtured with absolute pure love. All of creation was divided into two major divisions; heaven and earth.

Heaven is all of the spiritual levels and dimensions which are given to three main personalities: **1.** Powers, entities that have ability to control, direct and manipulate the forces from the One GOD. **2.** Principalities, spirit-born sons and daughters of God that are creators, administrators, and controllers in authority over all of creation. **3.** Angelic host, the servants of the One GOD being made manifest.

Earth, the elemental universe, is wherein new life is formed or manifested and given the opportunity to join with the One GOD in His great plan of creation. God is spirit, we are elemental and we have, by choice, the ability to become one with God. Scientist say all that was a natural outcome or an accident. We should ask from what, Nothing or God? If it came from nothing then it will return to nothing, all that has been learned will have been for nothing. If it came from God then it will return to God, and all that was experienced and learned will be retained.

--- THE ONE GOD-- ---------

THE BESTOWAL

The first step of creation called for the One GOD to bestow or give Himself, His total being, in order to bring forth those great plans. Without hesitation, it was done.

The One GOD is so secure in Himself that He can break Himself down and reconstruct Himself in any way He desires without losing a single fragment of His total being. The One GOD uses those forces and powers from within Himself to create all of creation, thus all creation came from God. We exist within that creation, thus we exist in God. There is a portion or fragment, like a cell from the body of the One GOD in us, thus God exists in us. He is the father of His creation. He is our Father. We are all His sons and daughters. Thus, we are all brothers and sisters.

A SIMPLE STORY OF CREATION:

The One GOD finished the plans.. The One GOD projected a great portion of power out from Himself which is known as Paradise. The One God projected His total being into Paradise, we know this as God the Father, the will of the One GOD being made manifest. The Father called forth another. This we know as the Only Son, the pure unconditional love of the One GOD. The Son turned to God the Father and swore absolute loyalty to Him and the plans of creation of the One GOD. Immediately there came forth another, the Eternal Spirit, the creative abilities of the One GOD being made manifest. The Eternal Spirit reviewed the Father then the Son and immediately swore absolute loyalty to the Father and concurred to being a conjoint worker with the Son in bringing about the Great plan of creation.

Understanding God's Plan Of Creation

The word Ineffable means one that is so magnificent, so glorious, so wonderful and so powerful that if we knew its name, we could not speak it. God is ineffable. **Omni** means the all perfect principal of whatever follows after it. Thus Omni-Omni means the all perfect principal of all perfect principals. God is Omni-Omni. God is **Spirit.** That is the **Light** of **Life,** which is a homogeneous power and force found only in the One God and Paradise. It starts out as one, and as it extends or travels out from the One, it separates and becomes two or more powers or forces that are different than it was when it started. Science has been looking for a connection that joins everything together. They haven't found it yet, but when they do, will they be able to detect the Life forces in it? We spirit-minded people have faith that it is there, even though we don't understand it at all. Or will the scientist find only the outer perimeters of another dimension of the One? We can learn a lot by asking the simple question, what did the One God use to create all creation? We are not just talking about the creation of the elemental universe, but, all of the creation of spirit, matter, and all other aspects such as multi-universes, parallel dimensions, and the holograph universe, the appearance of creation from nothing. The answer becomes self-evident when looking for WHY. He, the One God, used the force, the power from within himself. Thus all creation came from the One. That is the breaking down and reconstructing of the One, the bestowal, or giving Himself. We exist in the multi-dimensional creation that came from the One. Thus we exist in the One. We are experimentally learning about God's great creation.

God Divided Creation Into Three Divisions:

1. Paradise: This is the first singular heaven spoken of in Gen. 1:1 of the K.J. Bible it is the Absolute pure spirit from the One God. It is stable. Time and space do not exist in paradise. Travel is volitional in Paradise. It is the only place the monolithic force of all creation is found. It is the powerhouse of creation. It is the permanent abode of God the Father. There is a deity present in the spiritual gravity that radiates around Paradise and extends out through creation into and throughout outer space. It is the motivating force driving all of the creation. That is known to science as dark energy, even though it is a light that is so brilliant that we cannot detect it. In truth, it is from whence each of us came. We are a fragment, a seed a portion of the spirit of the One God via God the Father. Our true nature is a spirit. It is wherein we are taught the highest spirituality and develop to our highest spiritual potential, via the ascension process of becoming one with and companion to the One God.

2. The Central Heaven: This was created by the Only Son and his conjoint worker the Eternal Spirit. It too has its deity present in the spiritual gravity which radiates throughout creation wherein the life forces exist. The central heaven was created in a completed state with complete harmony and balance. It is the higher spiritual dimension where we progress after being recognized as true ascending sons of God. It has authority and dominion over the third heaven.

3. **The Heaven of heavens:** Consists of one spiritual heaven with seven heavens circling inside it. Each of the seven heavens is created by a triune of Ancient of Days. Each of the seven heavens is made up of billions of local heavens or universes we call galaxies. The 3rd. Heaven is made up of spiritual energy (dark matter) and elemental matter (two different dimensions). The physical elemental worlds function within the elemental gravity force. They are known as the schools of hard knocks and is called the holographic worlds. We learn from our experience in the mortal flesh and progress into the morontia being (Spirit and Matter) to progressively become spiritual personalities. This physical universe is where we learn by choice and experience. Some people call this a holographic universe. It is not real; unless of course you are caught up in it, then it is very real to you.

All of the creation was conceived and planned out with absolute pure love and perfection, nothing was left to chance, but a lot was left to CHOICE. All free-will creatures have the God-given right to choose to work with God in His great plan of creation. By our choices, we show our intent.

With each choice, there come responsibilities and consequences. By each moral choice that is in keeping with the will of God is how we progress inward or upward through the many different levels, spheres, and dimensions of God's wonderful creation.

God is a perfect balance of love, compassion, and mercy, with absolute justice. What love, compassion, and mercy cannot win over, justice will annihilate and reuse. Love ye one another, the one lesson that all worlds must first learn. First, we fear, then we learn to respect then we learn to love. The way of the kingdom of heaven is LOVE. The way into the kingdom of heaven is SERVICE to and for one another.

The One God projected forth all the power and force to create all creation. He brought forth powers, spiritual personalities that have the ability to funnel, direct, maneuver, control, maintain and manipulate the awesome powers of the One God. They are directing it throughout all levels and spheres of creation. The One God brought forth principalities, sons and daughters of God that are given creative, administrative and authority to create vast creations from that force, The One God, through the sons, brought forth the entire angelic heavenly host to help in that creation. All those called forth; created from or manifested from the One God were those perfect personalities. Each was given all that was needed to accomplish its mission. If any combination of the Holy Triune desired to bring forth any number of entities for a particular mission having any number of capabilities they would hold those thoughts in their mind as those entities would appear before them as full-grown personalities. They would have all the qualities and attributes as required. These perfect creations were those called the first light in Genesis I. They were energy called into motion as kinetic energy. The remaining spiritual energy was potential energy waiting to be called into existence. That was the separating of the light from the darkness.

We as physical human beings are at the bottom of the totem pole of free-will creatures, (let us rejoice because we are on the totem pole.) Our free will is freely given to us by the One God; it was ours for the taking. In truth, it is the only thing that we have of any value to give to God. It must be given freely of our volition. We are not alone because when we made our first moral choice to do something because it was the right thing to do, there came forth un-ceremonially a fragment like a cell from the One God, a portion of the Father to indwell in our mind. It is often portrayed as a guardian angel, but a guardian angel functions differently. The guardian angel works outside the mind with other angels and heavenly host creating situations, sometimes mischievously, wherein we are provided an opportunity to make choices. The thought adjuster, that portion of the One God, works in our mind encouraging us to make the right choices. When we make the right choice, the angels rejoice by singing and dancing. When we make the wrong choice, they often weep for us. Truly the angels of heaven are our truest friends and love us dearly. But, none is truer and stay with us longer than our eternal thought adjuster, that precious portion of the One God that will stay with us and guide us through all the levels and spheres of God's magnificent creation.

We, as spiritual entities, came from the one monolithic force, God. When we decide to act with free will, which is a mental act, we claimed our soul, which is seminomas with our free will. Therefore, they (the soul and the mind) are companions depending on each other. Thinking the thought is an idle thought until we apply our free will, then it becomes an activated thought. Our soul has a complete recollection of all that we have experienced. God's spiritual energy is the life-giving force that permeates throughout all creation. When a channel is open, we choose and are granted permission from those spiritual entities in authority, to enter this dimension, this world, to experience this portion of creation. Once we make a decision, we are bound to it until we learn all that is required to graduate. Our choices determine the speed we learn and progress. There is a vast number of the heavenly host appointed to help, guide, and protect us without interfering with our free will.

The Four Cornerstones

Of God's Creation

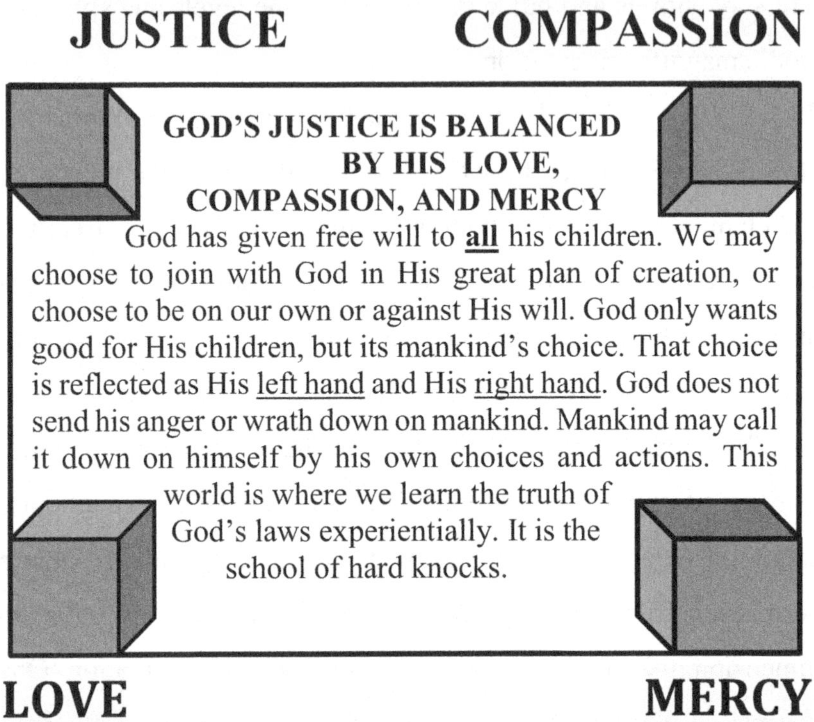

God's Justice is absolute and instantaneous. It is counterbalanced by his love, compassion, and mercy. If, while under His instantaneous justice, we sought to steal something or hurt someone, our hand would fall off and turn to dust before it hit the ground. God has given free will to all His children. Knowing they would make the wrong choices, He has tempered His justice with love, compassion, and mercy. That has the effect of His **left hand** and His **right hand**. Thus, most religious teachings those choice, on the one hand, walk with God, keep His commandments and be under His hand of love compassion and mercy and receive His blessings and protection, OR on the other hand choose error, sin, and wrongdoing and fall under His hand of justice. Thus God gives each of us ample time to learn from our mistakes and repent, change our ways, make the right choices. The CHOICE IS OURS. With every choice, there are responsibilities and consequences which become our blessings or our demons that we ourself created. What love does not win over; justice will eventually annihilate.

The One God That Became Three
The Eternal Triune

the Father: After the great power of Paradise was projected forth The One God projected, via his willpower, His total essence into Paradise. Therein we can know the will of the One God. The Father then called forth another like unto himself.

God the Only Son: This second person of the Holy Triune was the only one

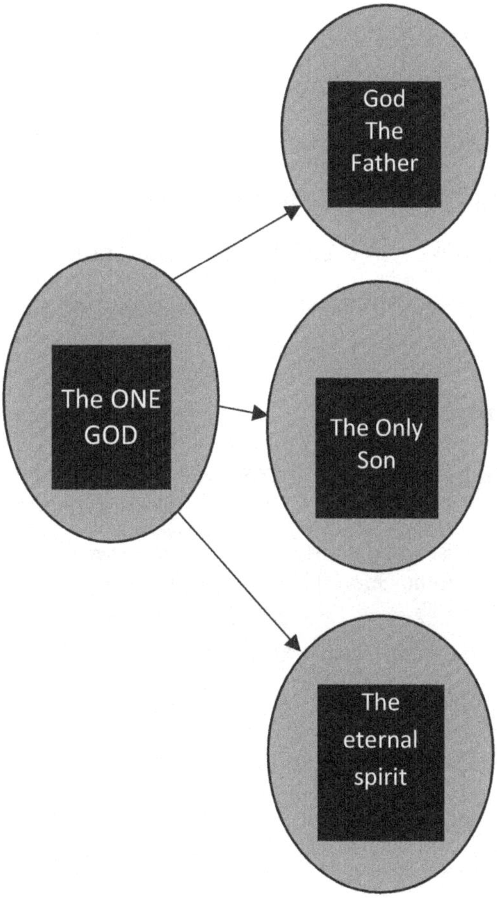

Called forth having the total essence of the One God like unto God the Father. It was an act of pure love. Thus we can come to know The Only Son by experiencing His unconditional love, the love of the One God. **God the Eternal Spirit:** When the Son swore absolute loyalty to God the Father, there was a manifestation, the coming forth from the One God, his creative wisdom and knowledge, the Eternal Spirit. As we allow the Holy Spirit (an exact copy of the Eternal Spirit) to become active in us, we can come to know the wisdom and knowledge of the One God.

THESE ARE THREE WAYS THE ONE GOD USED TO ESCAPE THE APPARENT ENTRAPMENT OF ABSOLUTE PERFECTION:

1: PROJECTING, **2:** CALLING FORTH 3: MANIFESTING

By learning about and experiencing God and his magnificent creation via the Eternal Spirit; and as we put into practice the pure love of the Only Son; and it becomes our hearts desire to do the will of God the Father; we will truly come to know and experience the One God.

Some of our ancestors came to understand the bestowal of the One God as projecting forth portions of His attributes. It can help to begin to understand His greatness. We can better understand the three of the Holy Triune as projections or manifestations of the attributes of the One.

THE ONE GOD

God the Father	God the Son	God the Eternal Spirit
Willpower of the One	Pure Love of the One	Wisdom and Understanding of the One
The Soul of the One	The Spirit of the One	the Mind of the One
Justice of the One	The Mercy of the One	The Compassion of the One
The Thought of the One	The Word of the One	The Action of the One

The purpose of creation was and is, to create an environment wherein new life can be born and grow experientially, and by their free will choose to become one with the One God in His great plans for creation. The primary purpose of the Holy Triune throughout all creation was and is to reveal the One God to that creation.

The Eternal Isle Of Paradise

In the beginning, God created heaven --- (singular) KJV 1:1.

Paradise is the eternal center of the universe of universes. The Universal Father resides at the very center of the upper surface of Paradise. The Eternal Sons personal presence surrounds the Eternal Father, and they are both invested by the Infinite Spirit. They then are surrounded by their divine coordinates and associates. The central Isle of Paradise is a gigantic organized body of cosmic reality in the master universe of material and spirit. It is from eternity and is the nuclear Isle of Light and Life. It is a stationary Isle that is composed of a single homogeneous organization of space potency not found elsewhere in the wide universe of universes. Jen> 1:3, God divided the light from the darkness.

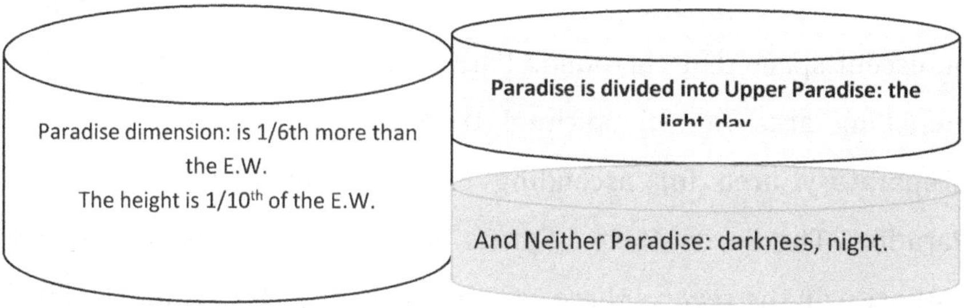

If the E.W. dimension is	50,	These are factors that various proportions can work out.
The N.S. dimension would be.	60	They could be miles, light years, or trillions of light-years.
The height would be............	5	They could be inches or ½", ¼," to draw a scale drawing.

Upper Paradise is divided into **three grand Spheres** **- DEFINED**

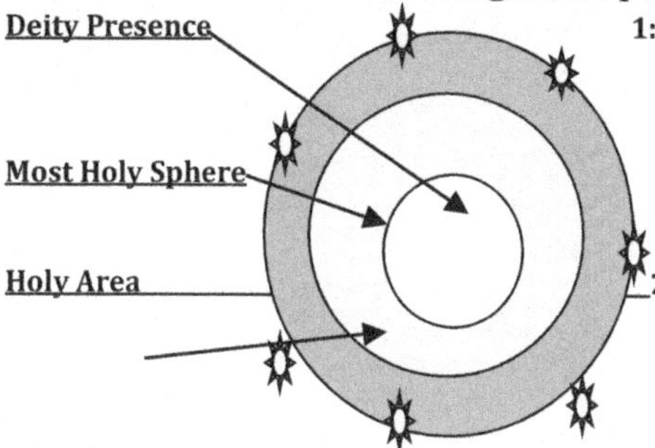

1: Deity presence: is the permanent abode of God the Father. It is so sacred that nothing physical can exist there. but it's spiritual glory is so awesome words cannot express it's beauty.

2: Most Holy Sphere: Is a wholly spiritual dimension or realm reserved for worship, trinitization and high spiritual attainment.

3: Holy Area: is an outlying residential region divided into 7 concentric zones sometimes called the Fathers mansions. The first three zones are reserved for paradise residents of Havona.

Nether Paradise

Is just underneath Deity Present. It appears to be the **HEART** of all creation. All physical force, and energy Proceeds from Paradise and will eventually return upon completion of the native circuit. Like our heart pumps blood through our bodies and returns to the heart.

Peripheral Paradise

Peripheral Paradise Is the third part of Paradise. It is a static quiescent space that surrounds Paradise. It serves as a landing and launching area for all spiritual travelers and serves as a final preparatory area for ascending entities before entering Upper Paradise. The Seven Master Spirits have their seats of power and authority on the seven spheres of spirit which circles about paradise. Here the Seven Supreme Power Directors slowly circles and dispatches certain paradise energies through the seven flash stations to the seven Heavens (or super universes)

THE ONE GOD

THE HOLY TRIUNE

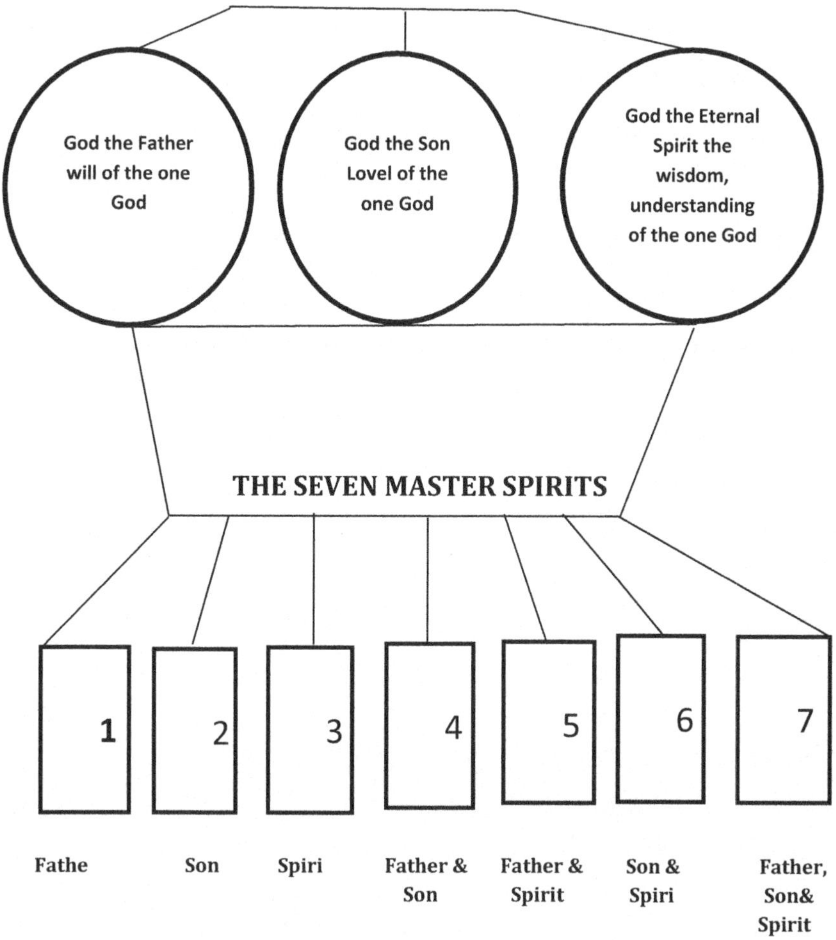

These are the maximum combinations of the three great expressions of the One God.

The 7 Master Spirits of God that sits around the throne of God.

The 1, the 3, and 7 combined = (11) constitutes THE THRONE OF GOD

These are those referred to in Rev. 4:4 & 5:6 and are the 7 fold spirits or 7 lamps before (roundabout) the throne of God. Seven is the maximum numerical expressions or combinations of the three of the Holy Triune. These Master Spirits directed, and still direct, the awesome spiritual power emanating out from the One God into the 7 Super Universes. Each Super Universe is governed and reigned by a triune (3) of Ancients of Days. We exist in the 7th Super Universe in the heaven of heavens called the firmament. In Daniel 8:13-14. One like unto the Son of man, Christ Michael, Jesus, went before the Ancient of Days and received his kingdom, our local universe, our galaxy.

Maltese Cross of Paradise

The respiratory system of creation

east/west dimension as space moves in from the N & S direction.

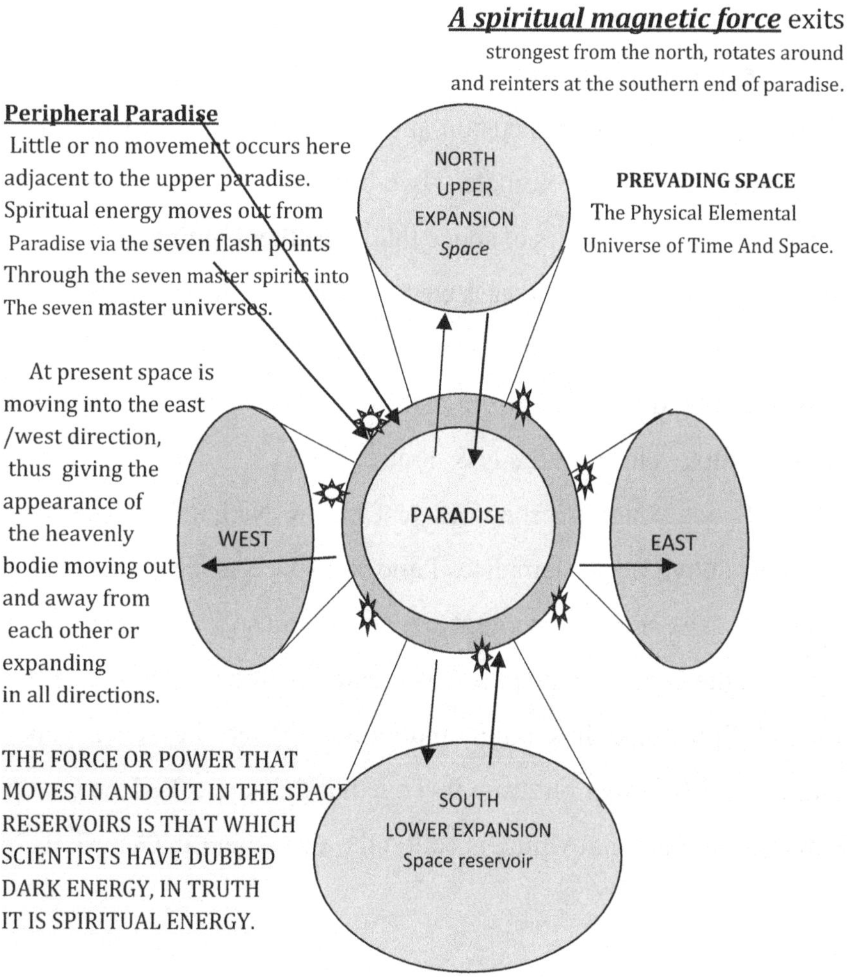

That movement involves the movement of space alternately expanding and contracting. There are two space reservoirs. One above and one below (N & S) of that which we call creation. They are connected to the Isle of Paradise. As they contract (squeeze in), it causes space energy to moves around paradise and into creation (an E & W movement). That contraction causes all creation to expand out and away from each other. This would give some support to the closed universe theory. The third movement is constant movement in all directions simultaneously. It is said that there are a billion earth years' of expansion and a billion years of contraction, making a cycle of approximately two billion years. The north/south expansions are reservoirs of space that are either coming in or going out, around paradise. The east/west direction is the location of the spiritual and elemental universe. Imagine if you will our total universe extending far out from paradise on the east/west dimension as space moves in from the N & S direction.

Space roughly originates just below Nether Paradise; and time just above Upper Paradise. Time and space as we understand it does not exist in Paradise. Out beyond Paradise is the location of Havona, the central heaven. It was created by the Only Son and the Eternal Spirit and has reign and domain over the seven super universes still further out from it. Together they are called the Grand Universe and are that which is called the firmament in Genesis 1:7.

The Isle of Paradise

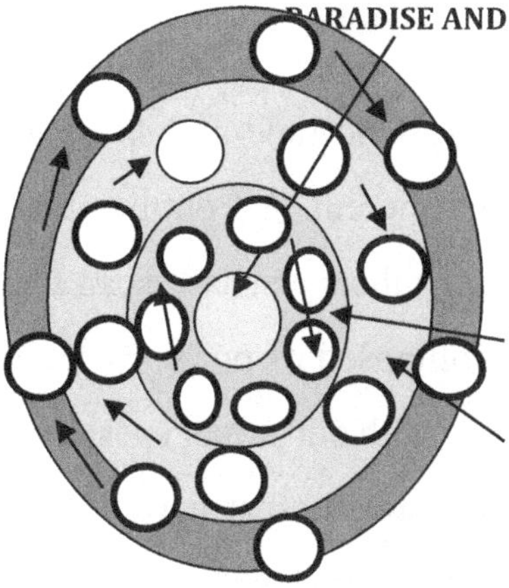

PARADISE AND THE THREE SACRED SPHERES OF PARADISE

the 1st Heaven: **Upper Paradise**; the light, kinetic energy called into motion, Deity personality residence.

Nether Paradise; darkness, non-personal energy, and power. It is potential energy waiting to be called into motion.

Peripheral Paradise; a quiescent space that surrounding paradise.

HAVONA, the 2nd or Central Heaven: begins just outside Nether Paradise and consist of the three inner spheres, called the 3 spheres of Paradise, because of their function, but they are a part Of the Central Heaven, which also consists of seven additional spheres. All 10 spheres circle around Paradise in a clockwise motion. The 3 inner spheres have 7 worlds in each, while the 7 Havona circuits have over one billion highly spiritualized worlds.

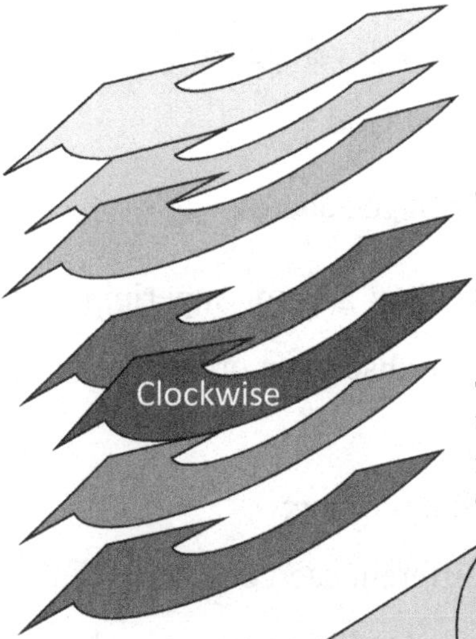

TWO GREAT TUBULAR BELTS OF AN UN-BELIEVABLE NUMBER OF MASSIVE DARK BODIES SURROUNDING PARADISE & HAVONA.

The inner belt is concentric circles. The outer belt is 50,000 times higher than wide. If the inner circle was 1 thousand ??? The outer circle is 10 Thousand times taller.

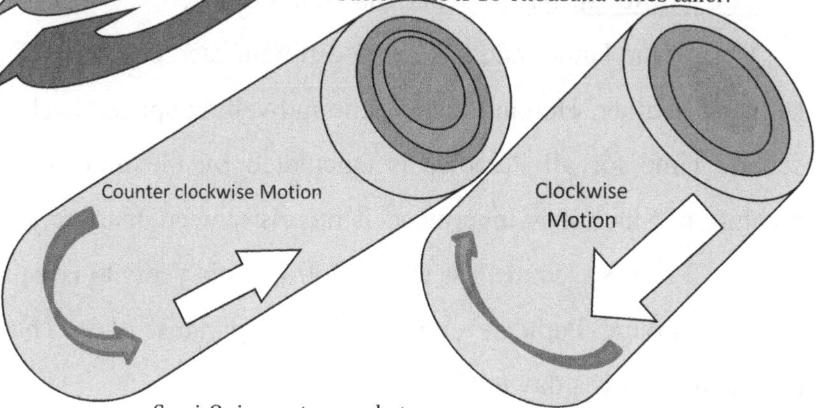

Semi-Quiescent space between All 10 spheres, the 2 cylinders of Dark Matter completely circle the Isle of Paradise.

The Isle of Paradise

Is the absolute center of all creation. It is stationary, and all movement therein is solely volitional. Entities will decide for themselves where they want to go. There is no time and space in Paradise. Physical elements cannot exist therein. All entities that reach this level of spirituality will perceive others just as we can see people of this realm. Their bodies are composed of purer spiritual essence. We, therefore, could not see them with our carnal eyes. The magnitude, beauty, and glory of Paradise are so great it would be impossible to begin to imagine it from a physical point of view. Thus we will have to wait until we are spiritually developed and can enter in to experience the truly glorious glory of God.

The Central Heaven Consist Of 10 Spheres

The 10 spheres encircle Paradise. They are not superimposed one over the other; they whirl around paradise in an orderly fashion in a clockwise direction.

The 3 inner spheres are paradise functioning:

1st is of the Father and has seven different sacred spheres of the Universal Father. Herein is where the indwelling spirits dwell and depart. Time for all Paradise is calculated by the inner circle's revolution. Since it is innermost, it travels slower. It takes just 7 minutes 3 & 1/8 seconds less than 1,000 Urantia years to complete an orbit around Paradise and is called a paradise day. Thus a thousand years is a day to God.

2nd. Is of the Son and has seven reflective spheres of worlds, all of which are identical. They act as reflectors for the pure spiritual energy from the Father God throughout Havona and are directed by the 7 Master Spirits out into the seven super universes of the Third Heaven, the Heaven of Heavens

3rd It Is of the Eternal Spirit. It too consists of seven different immense spheres and is the headquarters of the Seven Master Spirits. There is an immense amount of activities in this sphere. The seven remaining spheres of Havona also move in a clockwise motion and are the first steps of entry into paradise. We will progress from the outermost to the innermost sphere. Each step will bring about a more intense spirituality and attunement to the will and love of the Father God. Our bodies will be more and more spiritual, and our minds will become more acutely focused on the One God.

The Two Great Belts Of Dark Bodies

The dark gravity bodies are unlike other space bodies in that they do not absorb or reflect light. They are non-reactive to physical energy light. It is spiritual light, a non-heat light, unlike physical energy light that also gives off heat. They so completely encircle Havona as to hide it from even a nearby inhibitor of time and space. The outer belt is 10 thousand times higher than the inner belt. The up and down arrangement of the dark bodies of the outer belt is 50 thousand times greater than the transverse, or across. It appears that the pure spiritual energy of the One God is generated in the heart of Nether Paradise and is projected out by God the Father and is reflected off the seven reflective worlds of the 2nd Paradise Sphere of Havona and out and through **the first belt** of dark bodies. Passing through the 1st belt puts the energy into a wave motion which would be of equal wavelengths. Then as it passes through **the outer belt**, it is made to spread out into different frequencies, a stepping down of the awesome spiritual power of the One God, somewhat like passing white light through a prism, it comes out as the seven colors of the rainbow. There is also created around paradise a spirit gravity that does not pass through paradise but around paradise through the quiescent sphere of Peripheral Paradise. It enters at the southern end and exits at the northern end of Paradise. Somewhat like our physical gravity field protects our planet from most of the sun's rays. This spirit gravity is not responsive to physical energy, light or gravity, but all creation is responsive to the spirit gravity. God the Father radiates, or projects, the spirit power out from the One God, out from Paradise. The Son draws that spirit energy back toward Paradise in the form of spirit gravity. To say that another way, the One God, of his free will, projects himself out as fragments, via God the Father. Those fragments dwell in the new living creatures. The pure love of the One God (the Only Son), acts as a magnet pulling that spiritual energy back to Paradis. That love brings with it those new evolving and ascending entities, you and I. That spirit gravity, the pure love of the One God feels like an emptiness in our heart, our mind, and our life that only the love of God can fill. God's love is not restricted to gender, religion, group, nor world or dimension. It is sent to all that will respond to that love.

God's pure Spiritual Energy changing frequencies as it passes through the two belts of dark bodies.

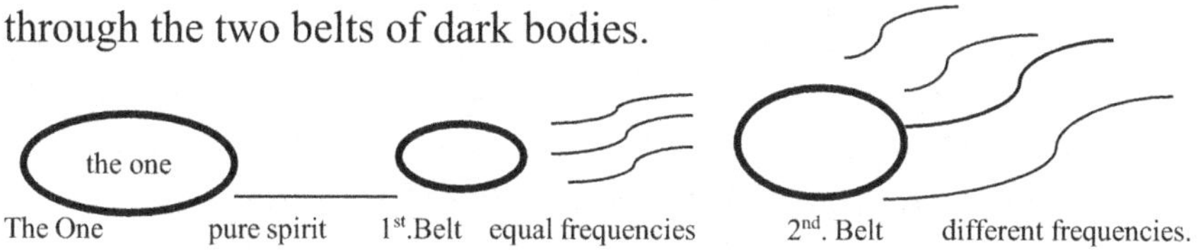

HAVONA THE CENTRAL HEAVEN

7 SUPER UNIVERSES OF SPIRIT AND MATTER REVOLVE AROUND INSIDE THE 3RD HEAVEN, AND TOGETHER THEY ARE REFERRED TO AS THE HEAVEN OF HEAVENS. THIS HEAVEN OF HEAVENS) (THE 3RD HEAVEN AND HAVONA, THE CENTRAL HEAVEN (THE 2ND HEAVEN) REVOLVE AROUND PARADISE (THE 1ST HEAVEN).

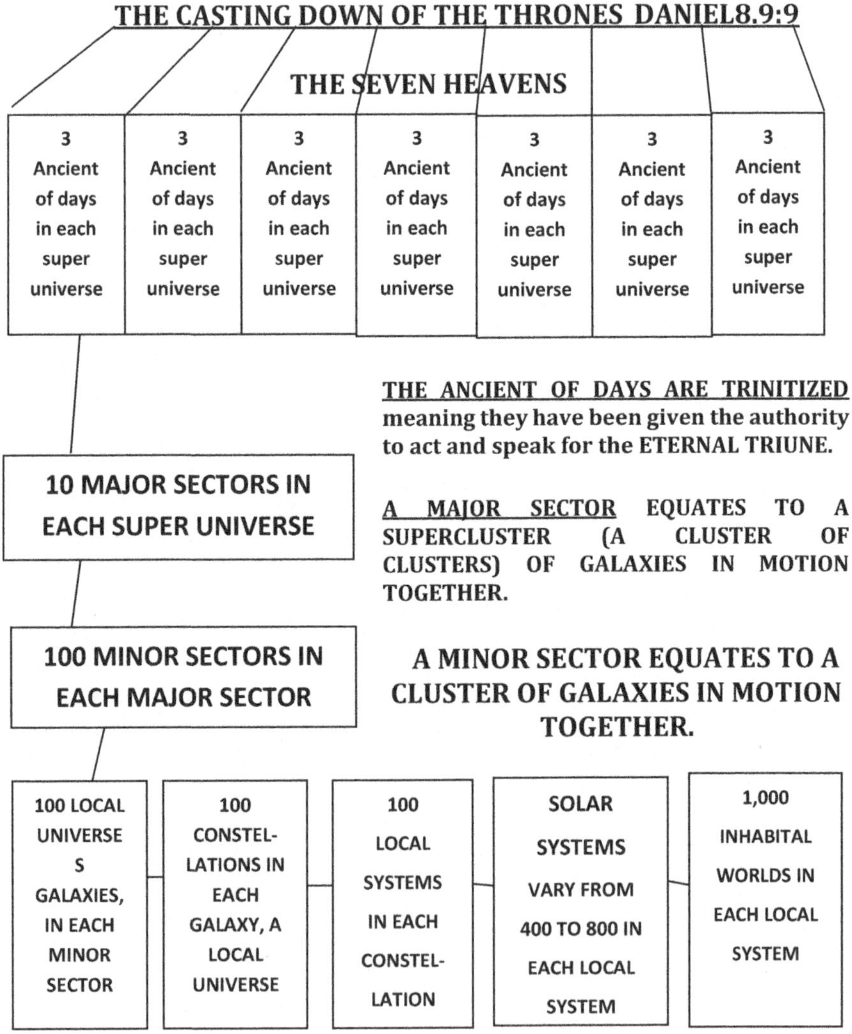

THE ANCIENT OF DAYS ARE TRINITIZED meaning they have been given the authority to act and speak for the ETERNAL TRIUNE.

A MAJOR SECTOR EQUATES TO A SUPERCLUSTER (A CLUSTER OF CLUSTERS) OF GALAXIES IN MOTION TOGETHER.

A MINOR SECTOR EQUATES TO A CLUSTER OF GALAXIES IN MOTION TOGETHER.

There is a SPIRIT born creator son of God, referred to as a Great Prince, in charge of each local universe, or galaxy. When our, Great Prince is perceived in the spirit, he is often referred to as being like unto the **"son of man,"** when perceived in the physical he is often referred to as

Stepping Down Of The Eternal Triune

The Holy Triune steps their powers down as the **Seven Master Spirits** which are expressions of **1**: The Father **2**: The Son **3**: The Eternal Spirit **4**: Father & Son **5**: Father & Spirit **6**: Son & Spirit **7**: Father, Son & Spirit. The Seven Master Spirits are the great directors of the tremendous power of the One God directing or funneling that power into seven great heavens within the firmament, which is the Heaven of Heavens. Thus the numbers One, Three, and seven are significant in creation. The Holy Triune is further stepped down as the **trinitized Ancient of Days**, the THREE directors and supervisors of the SEVEN Super Universes. The next step down is a trinitized **perfection of Days**; they are the supervisors of major sectors. **Trinitized Recent of Days** are supervisors of Minor Sectors. Local Universes, or Galaxies, are created and governed by **a Michael Creator Son** and has **a Trinitized Union of Days** as counselors and advisors. This position in our Local Universe is held by Emanuel, the beloved friend of Christ Michael. Each Constellation has a **trinitized Faithful of Days** as advisors, and they are the furthest expression from the One as the Holy Triune. There is a portion of God the Father that dwells in each of us. It is a complete expression of the One, like a cell from the body of the One God or a mustard seed. It enables us, the lowest creature on the totem pole of free-will creatures, to communicate directly with the One God. By projection of the various attributes and qualities, the One God can be a part of all creation from the absolute to the finite.

In Genesis I: 6, God created the firmament, which is an expansion, and divided the waters, (which is a word symbol for spirit,) from the waters and God called it the firmament or heaven. Heaven also refers to a spiritual dimension or dimensions. **Verse 9:** God made the dry land appear, not just our planet, but the entire elemental universe in the lower part of the firmament. God thought the thought; the Only Son spoke the words and the Eternal Spirit along with all the heavenly host, those that are called the light,(the first light of **Genesis 1:3**) went to work, marking the beginning of creation. First: the heavens were created with spiritual energy and then with spirit-matter (Dark Matter), and then the elemental cosmos. That began in the outer perimeter of Paradise because Paradise is eternal as the One God is eternal because it came forth from the One God. Paradise is the permanent abode of God the Father.

The Only Son and the Eternal Spirit conjointly created the Central Heaven. It was absolute perfection in a completed state of spiritual energy (dark energy). It did not evolve. It is not of the elemental universe. The Central Heaven is the abode of the only Son and the Eternal Spirit. The Central Heaven reigns over and has authority over the Heaven of Heavens, the third heaven.

As spiritual energy steps down (or condenses), it takes on more substance, and it becomes spiritual matter (dark matter) the heaven of heavens is made up of spirit-matter and elemental matter. The elemental matter is the cosmos as we know it, that which we can measure and calculate, or at least most of it. It was created from spirit matter as it (spirit-matter) steps down or condenses or takes on more substance and becomes our cosmos. The central heaven and the heaven of heavens constitutes the Grand Universe.

The Physical Elemental Cosmos

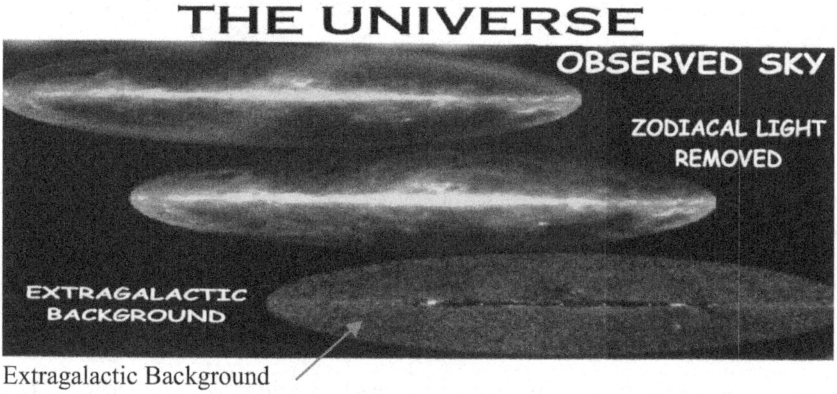

Extragalactic Background
All-Sky Maps of the Infrared Background credit to NASA/COBE/M. Hauser

Note: the center of the extragalactic background above. It appears to have an area that is not physically active. If it were spiritual energy, it would not be visible. Perhaps one day, our scientific brothers will discover a way to see what lays within. Some say there appears to be a gigantic black hole in the center of the universe. Perhaps, it may well be the outer perimeter of the outer belt of those black magnetic bodies surrounding Havona (The central heaven) and Paradise.

GENESIS 1:8 THE SECOND DAY

Coming Forth Of The Seven Super Universes

Within The 3rd. Heaven

The seven master spirits directed and funneled the awesome power of the One God into seven different areas of the firmament, causing seven super universes to begin to materialize in the area labeled the extragalactic background above. Imagine there being seven great revolving masses in a counter-clockwise motion inside that area. Scientists are beginning to detect some of those separations already. Perhaps they will, in time, discover a better perspective and be able to distinguish the heavens within that heaven, that which Solomon called the heaven of heavens. When the spiritual power directors had made those forces ready, the Ancient of Days came forth and assumed the position of authority as was seen by Daniel 8:9. Remember, the Ancient of Days is three trinitized spiritual personalities who work conjointly as one and often speak for the One God in the first person. Later in Daniel 8: 13 , Daniel saw one like the Son of Man, a title that Jesus took onto himself while incarnate in this world, which is a part of His kingdom. We know that kingdom as our galaxy in which we live.

Scientific Look At Our Universe

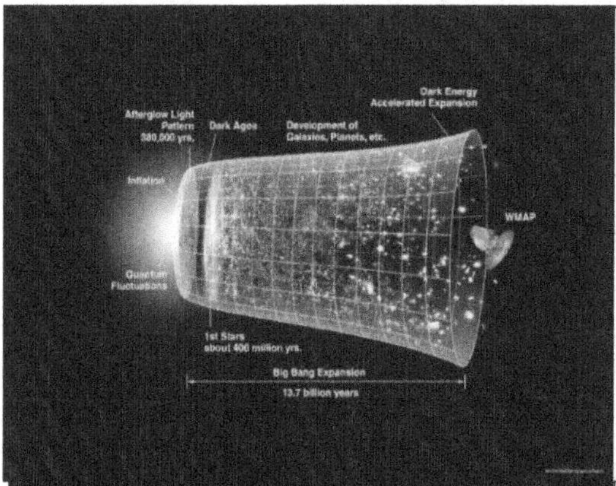

In 2001 NASA launched the WMAP probe to measure the microwave background. One of the many photos (L) has exploded mankind's awareness. This photo shows the growth of our elemental universe in astounding detail

CREDIT TO NASA/ WMAP

SCIENCE TEAM

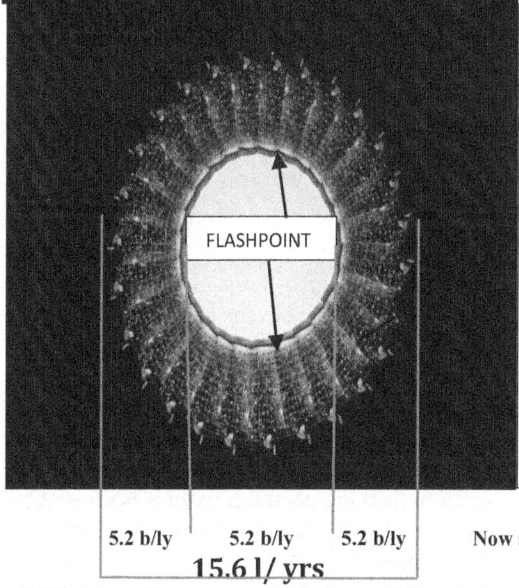

Since the probe was launched from our earth and is looking at the universe from that prospective, it gives a view like a slice of a pie.
We do not get a full view of what our universe may look like. By making a composite with 27 photos joined together will create an elliptical or concentric elemental universe around the point of inflation. (R) This gives a view of the elemental universe like, from above.

Now science is suggesting the dimension is 93 billion L/Y.

If this is true, then we can get a better or different view of the universe.

It would resemble the extragalactic background from pg. 41. That would depict a 13.7 billion light-years wide band surrounding the flashpoint which appears to be a 27.4 billion L. year center. That center is that which has been referred to as Paradise and the Central heaven, see pg. 44. The point of inflation would appear to have occurred around and outside of that center, marking the beginning of the elemental universe.

The Three Great Heavens

SOME CALL THEM THE THREE GREAT LIGHT REGIONS AND OTHERS CALLED THEM THE THREE GREAT MYSTERIES

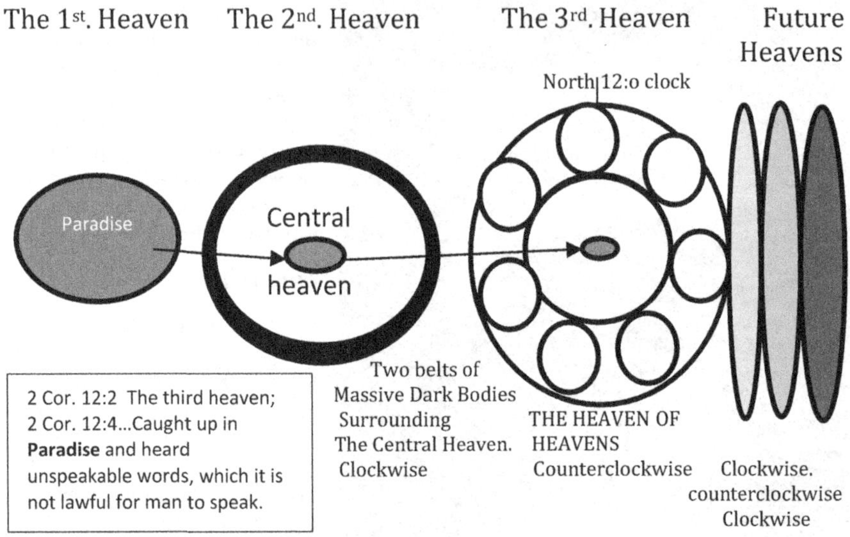

Deut. 10:14 Behold the heaven, and the heaven of heavens is the Lord's thy God. In 2 Chr. 2:6 Solomon said who can build him a house, seeing the heaven and the heaven of heavens cannot contain him, also the same in 1 King 8:27 Solomon said "... Behold the heaven and heaven of heavens cannot contain thee.

The one power, the primordial force, the One God projected forth a tremendous amount of brilliant, luminous spiritual light and energy, into which he willed (projected) himself and created the Isle of paradise making it his permanent abode. it is the throne of God. his deity presence creates a spiritual magnetic force around paradise and extends throughout creation, which is the driving force (dark energy) of all creation. then the central heaven was created by the only son and the eternal spirit. it was complete and perfect of the highest order of spirit energy, with its own deity presence spiritual gravity permeating throughout creation wherein life exists.

The seven master spirits projected seven great massive amounts of energy inside the third heaven. they revolve together around inside the central heaven. together they are called the grand universe and is the heaven of heavens. spiritual energy slowed, down which caused spirit matter (dark matter) to form. as dark matter slowed down, it formed the elemental universe, which became matter. each has its gravity forces and is reigned by the creator sons of God. the escaping energy is collected in three regions outside our present universe, which will probably, be used in future creations.

Expansion Of Creation

PARADISE: Stationary Living monolithic force, the center of creation.

CENTRAL HEAVEN: Stabilized spiritual energy is moving Clockwise around Paradise.

HEAVEN OF 7 HEAVENS Spiritual and elemental energy Expanding and Contracting Moves counterclockwise.

These three Constitute the --- GRAND UNIVERSE

APPROXIMATELY 13 TO 14 BILLION LIGHT YEARS

All living creatures or entities

EXIST IN OR LIVE IN THE GRAND UNIVERSE.

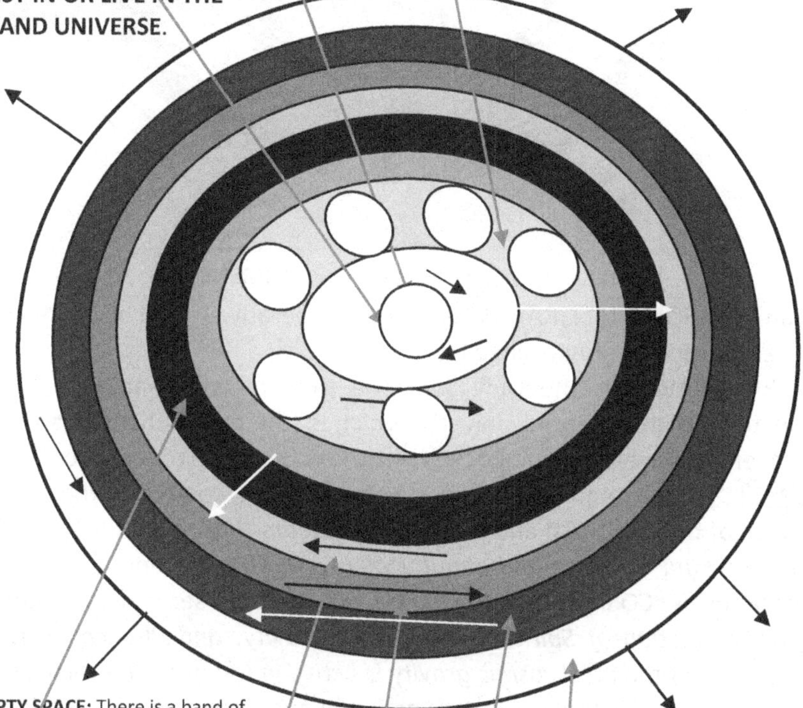

EMPTY SPACE: There is a band of 400 to 500 thousand light-years of space that is void of stardust-cosmic fog.

1st. Outer space level is a band of extremely active energy increasing in volume and intensity that expands out from the band of empty space for about 25 million light-years. It is rotating clockwise.

2nd. Outer space level: is more intensely energized force rotating counter-clockwise and is probably a presage, for the organization of more material creation.

3rd. Outer space level: Is a continuation of expanding energy forces rotating in a clockwise motion.

4th. Outer space level: Extends out and into infinity with no end. That would be more than 1 to 2 TRILLION light-years and growing.

Deity Presence Gravity

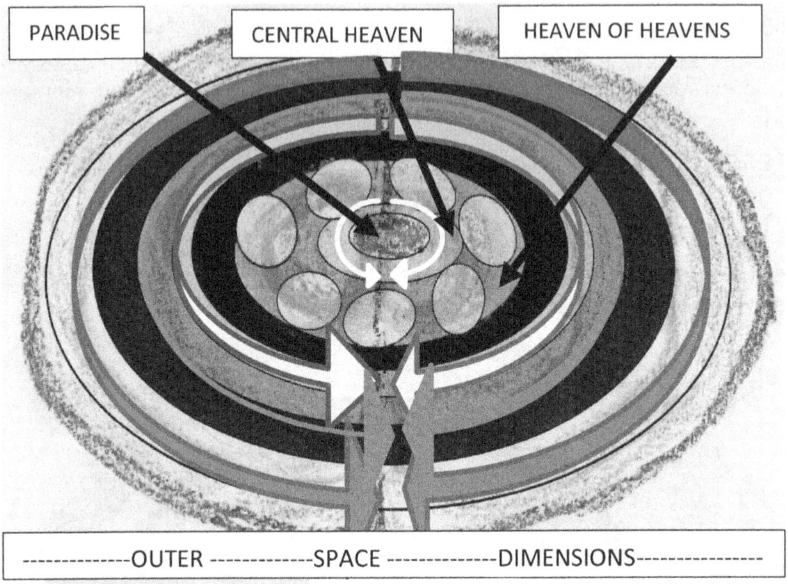

God's PERSONALITY GRAVITY is exclusive in paradise, but it also functions on all 4 absolute gravity circuits.

SPIRIT GRAVITY is the Only Son. The total of which appears to be functioning in the Grand Universe which is (the central heaven and the heaven of heavens combined) with little functioning in outer space. MIND GRAVITY is the conjoint Worker, the eternal spirit, combining the mental forces of both spiritual and physical. It *appears that about 85% is active in the Grand Universe and about 15% is probably working in the outer space levels.* COSMIC GRAVITY is the Isle of Paradise. *It is a combined gravity function of Spirit Gravity, Mind Gravity, and Physical Gravity. About 5% of the total cosmic gravity is active in the grand universe. That leaves about 95% that is functioning in the outer space levels.*

4 Absolute deity present gravity forces	Isle of Paradise powerhouse	Central heaven	Heaven of heavens wherein Life forces exist	outer space new creations
Father's personal gravity	100%	100 %	100 %	100 %
Son's Spirit Gravity		(99 %)		1 %
Spirit's Mind Gravity			(84 %)	15 %
Cosmic Gravity, Spirit Mind Gravity and Elemental Gravity.		10 %		90 %

God the Father, Son, and Eternal Spirit are multidimensional creators and are multifunctional entities, they can and do function on or at all levels and dimensions simultaneously, at the same time.

Parallel Dimensions

When we depart this physical elemental world through the process, we call death which is really a process of transiting the physical dimension into the Moreton dimension which is within the lower spiritual dimension of what we call heave. Therein we can reunite with others that preceded us. After a reasonable time, we may choose to remain with them and progress with them later, or we may be taken to the appropriate level that is best needed for our spiritual growth. There are many levels within this dimension, and we will meet many wonderful teachers and friends. Eventually, we will progress into the Local System wherein we will meet even more spiritual guides and teachers. Our progress will culminate as we come to know and love the system sovereign. This is where Lucifer was assigned before he started the war in heaven, which was terminated by Gabriel the Morning Star. Melchedick, the hi priest of the Most High God, was assigned to this position of high authority.

PARELL DIMENSIONS

Different dimensions occupy the same or near the same time / space zone. Different yet similar. As we progress upward (spiritually), we become less elemental and more spiritual until we become all spiritual.

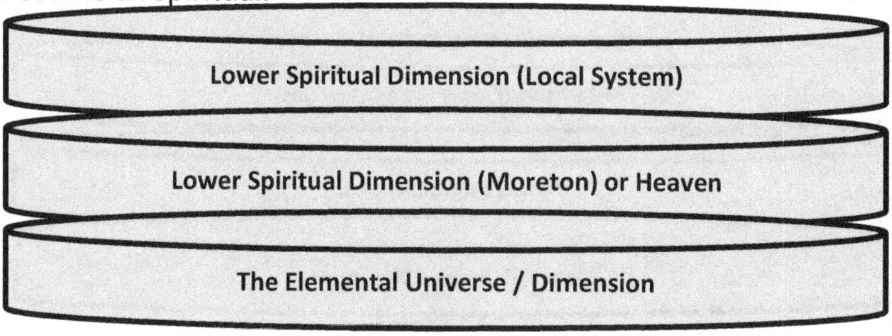

Progressive Dimensions

Our motivation should be to attain our highest spiritual level of enlightenment. The steps are laid out from the bottom (the elemental dimension) all the way up to attaining at one-ment with the One God.

It is a long, tedious, and sometimes difficult journey. The One God has given us an eternity to make that adventurous experiential journey. He has provided help and assistance every step of the way. The rewards for making the journey are beyond our comprehension and worthy of our every effort.

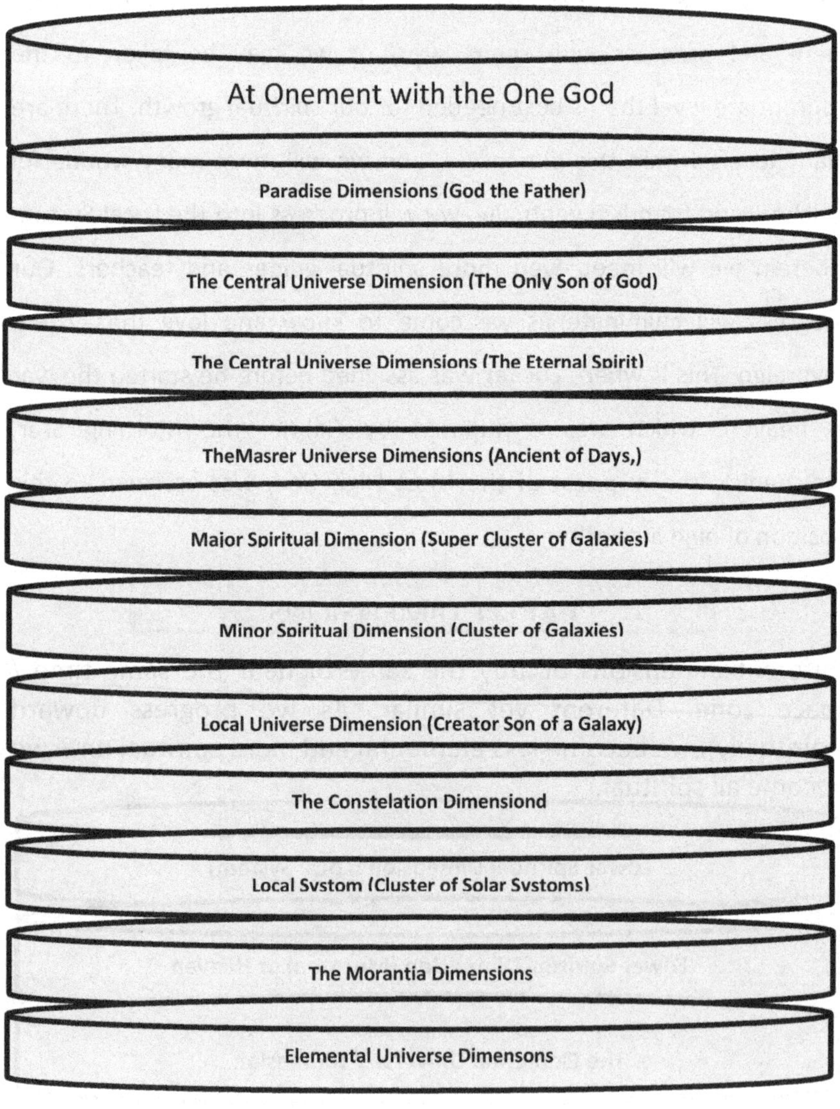

The Involution From Spirit To Matter

The purest form of God's spirit is called the monolithic force, it is pure and stable. It radiates out and steps down as it passes through the many heavens, levels, and dimensions. It first steps down into what is called pure spiritual energy (or dark energy) then into the higher spiritual realms of Spirit matter called dark matter which is a mixture of spirit and matter. It then steps down into our physical elemental cosmos and out into the outer space dimensions.

God first created Paradise then He, the Only Son and the Eternal Spirit called forth the heavenly host to create and administer to the remainder of creation. Along with them He called and projected forth all the sons and daughters, you and I. He, therefore, is our Father. As God is the father from which all creation came, the creator sons and daughters are our spiritual parents or creator father as we are the creator father and mother of our children. As creation began to unfold we as spiritual sons and daughters began to reach out and experience within that which was created, stepping down as we ventured out from our heavenly homes into the depth of this physical world which is but an illusion, a hologram, a school of hard knocks wherein we can learn and grow in wisdom, knowledge, and insight. We have chosen our path by the actions of our free will.

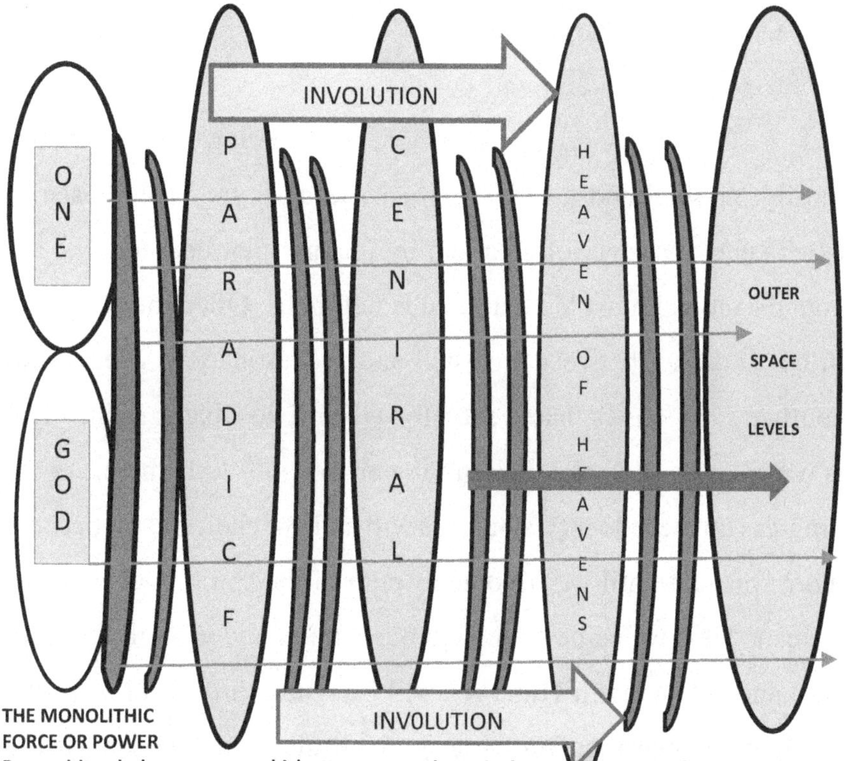

THE MONOLITHIC FORCE OR POWER
Pure spirit = dark energy which steps down And creates spirit-matter or dark matter which steps down and creates the elemental universe or the cosmos.

The Involution From Spit To Matter

All That Proceeds From God
Will Return To God In Its Assigned Time

The preceding pictorial representation shows how God's spiritual Omni-powerful energy steps down, but how does it return? We will look at two possible methods.

1. The spiritual/mental combination called consciousness.
2. Matter or energy.

1. CONSCIOUSNESS: On page 58 & 59 you will find a drawing of TOP/ DOWN OF CREATION If you look from the bottom up you will see in red spirit order 1 – 7 these are the seven major spiritual levels we go through as we spiritually and consciously ascend the spiritual ladder. At the lower level, we are mostly physical and a little spiritual. Because we have chosen to experience this level of creation, by making that decision, we are bound to it and must learn all that is before us. Often there appears to be a conflict between free will and predestination. There is no conflict. God knows that eventually, we will do what is right, when is our choice. We have chosen via our free will to be here. As we progress up the ladder (via our free will and our choices), we become more spiritual until we become as pure in the spiritual energies as we can. Thus we came from God our Father, and we will return to God and be one with Him. We will be rising through progressive levels and dimensions from a newborn physical consciousness to the awakening of a spiritual reality, becoming a wiser and closer companion with God. Consciousness is the vehicle of transition through dimensions.

2. MATTER: Our scientific friends tell us that energy cannot be destroyed; it can only be changed. Also, there appears to be a tug of war between gravity and friction. As gravity pulls matter in, it creates friction. Eventually, that friction overcomes the gravitational pull and explodes in a huge flash of light and expands outward until gravity again begins to draw it back toward its self. That works well for explaining a supernova and the like, but it does not hold true in the case of black holes. They act like a super monster gobbling up everything they trap, or like a vacuum sweeping up matter and energy which seems to dissipate. Or is it returning to spirit matter,- - too dark energy --and back to its source the monolithic force -the One God?

Let's apply the thought that everything came from the monolithic (One God) and will return to that from which it came.

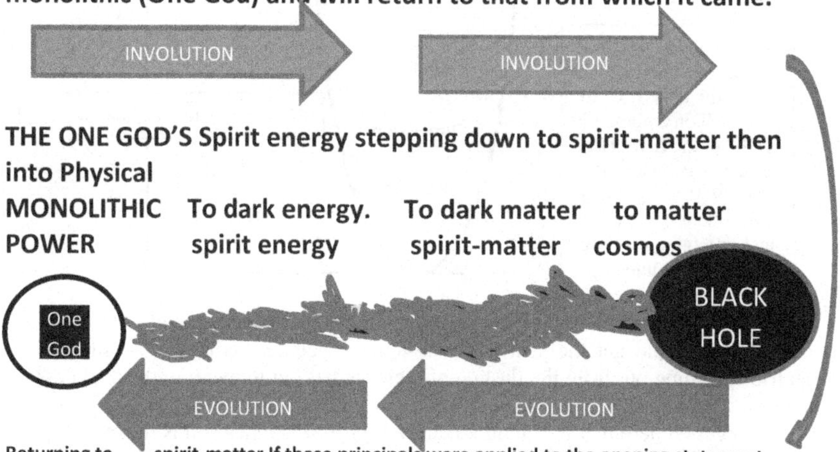

Returning to spirit-matter If these principals were applied to the opening statement that everything that came from God will return to God would work quite well.

Two Spiritual Dimensions

which reign over the elemental universe

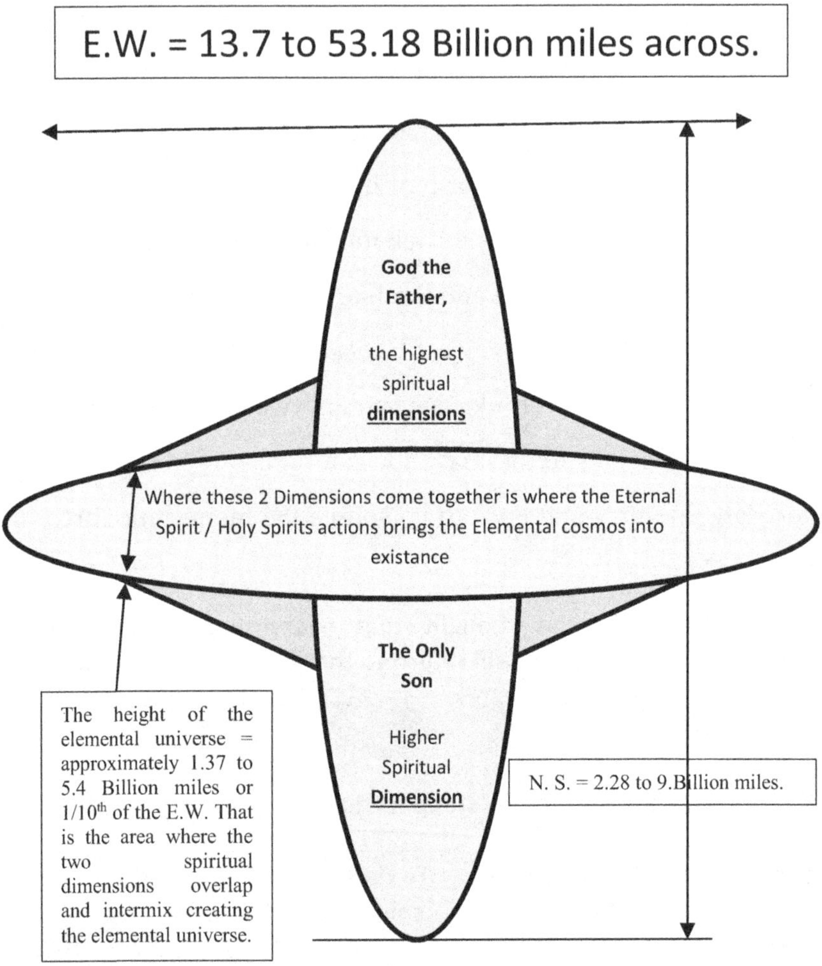

Science is suggesting that the universe, like a membrane, stretched, out in all directions. One would get the idea that it is long or square. Perhaps it is elliptical or concentric spreading out from the flashpoint. Science has yet to mention the thickness or depth of that membrane.

It is suggested here that it is approximately 1.37 billion miles and is the overlapping area of the two awesome spiritual powers or dimensions we call God the Father and the Only Son combine. The creative forces of the Eternal Spirit start the activities of stepping down or slowing down and combining those great forces molding them into that which we call creation.

God Is Spirit, And That Is Different Than Matter.

That homogeneous power we call spirit starts out as one force. As it goes out from paradise, it slows down and begins to separate. A process of stepping down from the purest spiritual energy or power becoming more substantial, or firmer, spiritual energy (dark energy) stepping down, slowing down and becoming spiritual matter (dark matter). It then appears to be two and more powers the further it goes from paradise. When God the Father and God the Son (who are both living spiritual energy). Where these spiritual entities Overlap their energies, the Eternal Spirit interacts with their spiritual forces and causes the energy to step down further, and the elemental universe powers begin to manifest in the lower or outer spiritual heaven which is called the firmament.

Since spirit is a different force or power, it functions with different gravitational laws. It is different from the elemental cosmos. Spirit gravity is not affected by elemental gravity, but it (spirit gravity) controls the elemental gravity. Spiritual energy is not visible to the elemental, but it creates the elemental universe. Both spirit and matter came into existence from the same one force, the monolithic force or power which we know as the One God and is conscious and is aware of being alive. If everything came from that monolithic force, and it did. If we are alive, and we are. Then that life force came from the One monolithic force. And it did. Therefore the monolithic force was alive with consciousness, and it was.

Top-Down Of Creation

THE ONE GOD
The One force, One power,
The One essence
Is the ineffable omni-omni essence from which All creation came, and in whom all creation exists.

PARADISE Oath of Finality 7
the 1st. Heaven Spirit Order 6
Came forth from the One.It is the permanent abode Spirit Order 5
Of God the Father. Finding the Father
Time and space begins beyond paradise
Inner Darkness,
The 2 belts of dark bodies that separate higher spiritual from the lower spiritual and the physical worlds.

plan of creation The bestowal of the One God.

Father of the microcosmic atom.

proton-	**neutron**	**Electron**
Father God	**the Son,**	**Eternal Spirit**
The will	**love**	**wisdom**
The thought	**the word**	**the creativity**

The One God is so secure in itself that it can break itself down and reconstruct it's self in any way it so desires without losing one fragment of His total being.
Absolute Pure Spirit And All That Are Brought Forth Here Are Eternal As the One God Is Eternal.
We must be transported by a Seraphim through this force field. The experience would be somewhat like that of death of the physical.

Para-
dise

Spirit Orders 1-4 and physical / morontia cannot exist in the pure spirit of Paradise.

HAVONA
The 2nd heaven and is the is the abode of the Only Son.
Grand Universe
central heaven, and the 7 super Universes Over which it reign.
THE HEAVEN OF HEAVENS

CENTRAL
HEAVEN
Spirit
order 4

THE CENTRAL HEAVEN
Was created by the Only Son, and his conjoint worker, the Eternal did not evolve. Those that pass Spirit. It was created in complete state and through it do or are evolving.
Master Universe
Includes 3 other un-inhabited undeveloped space outside the Heaven of Heavens herein new heavens may develop. Their motion is clockwise, counterclockwise, and clockwise.

THE 3rd. HEAVEN

There are 7 great heavens in motion inside the heaven of heavens. Each is ruled by 3 ancient of days, who are a direct representative of God the Father. We live in the 7th. Heaven and are now in a northerly path toward the fullness of time. Our heaven, the 7th. Heaven is Orvonton it is approximately 500,000 light years across.

North **12 o clock**

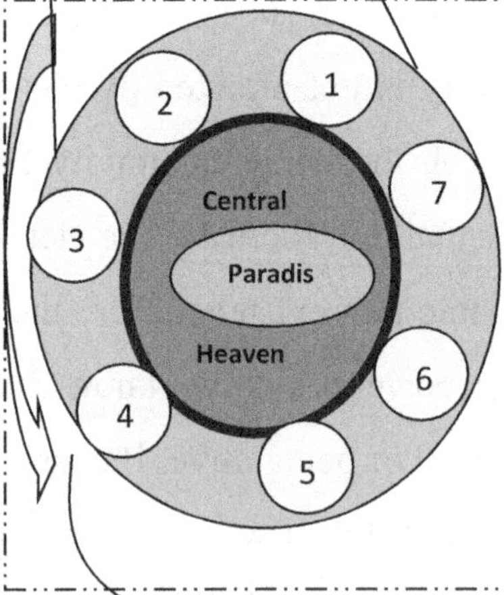

This is where heaven, the spiritual kingdoms, and earth, the elemental universe, come into contact. The heaven and heaven of heavens is that 15.6 trillion light year universe our scientists know about, the physical universe. Theology knows a lot about some of the spiritual heavens.

There is outer darkness that separates the heavens from the earth, spirit from matter.

Outer darkness is transcended by our morontia body at the time of death of our physical body.It cannot exist in spirit . Continued

ORVONTON Our super universe
The 7th. Heaven Uversa is its (headquarters)

Each of the 7 Heavens is made up of several layers of spirit, morontia (spirit & physical) and matter, the elemental dimension.

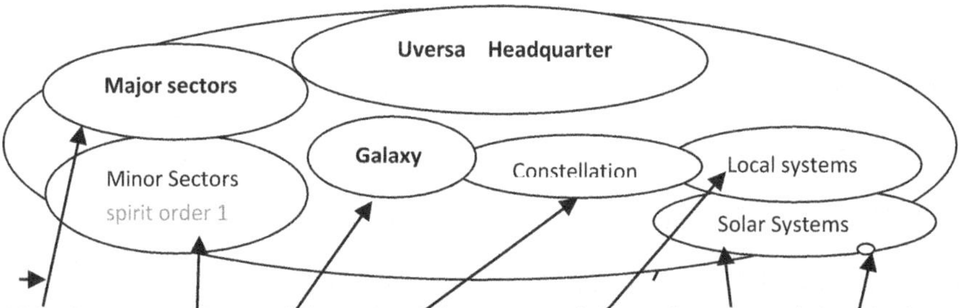

Major Sector	Minor Sector	Galaxy	Constellation.	Local syst.	Solar system	Inhabitable Planet
Splandon	Ensa	Nebadon	Norlatiadek	Satania	Monmatia	Urantia
Umajor 5 hq	Uminor 3 hq	Salvington hq	Edwntia hq	Jurusem hq	**500 +**	**Earth one of**
1 of 10 in Orv.	**100 in Maj.**	**100 in Min.**	**100 in Gal.**	**100 in Constl.**	**in local syst.**	**1,000 in loc. Sys.**
Orviton 10	**1,000**	**100,000**	**10,million**	**1,billion**		**1,trillion inhabitable**
7 Heavens 70	**7,000**	**700,000**	**70,million**	**7.billion**		**7,trillion worlds**

Overview Of Creation

The One God

Is that monolithic force or power with consciousness from which all creation came and is known as the One God.

Paradise

Is the 1st. Heaven (singular) referred to in Genesis 1:1 (KJB) which was projected forth from the One monolithic force, God, and then he projected His total essence into it and created the powerhouse of all creation, the Isle of Paradise. It is the permanent abode of the will power of the One God we know as God the Father. Paradise is the absolute center of creation and is stable (not in motion). Travel is volitional and instantaneously. All remaining creation has its origins from Paradise via the Eternal Triune, God the Father (the will of the One) the Only Son (the pure love of the

One) and the Eternal Spirit (the wisdom, knowledge, and creativity of the One).

The Light & Darkness

Refers to when The Eternal Triune called **all** the heavenly host (Powers, Principalities, and the Angelic Host) into being and gave them a purpose, existence, and reality (we could call it or them kinetic energy, the light). The remaining monomythic power is potential energy (darkness) waiting to be called into existence by creator sons and daughters (principalities).

The Central Heaven

Was created by the Only Son and His conjoint worker the Eternal Spirit. It did not evolve. It was created in a completed state and was in perfect harmony with Paradise. It rotates clockwise around Paradise and surrounding it.

Two Great Belts Of Dark Bodies

They were created. They neither absorb nor reflect light; they are spiritual energy and are not reactive to physical energy or light. They serve to isolate Paradise and the Central Heaven and to protect lower energies forces because the awesome power of the higher spirit would be harmful to lesser energies and entities. For a lower spiritual entity to transverse these two bands, that entity would have to go through a transition similar to that of the death of the physical. That body would be laid down, and a Seraphim would carry the entity into the Central Heaven it would have a new more rarefied purified spiritual body.

The Heaven Of Heavens

developed When the spiritual energy began to slow down and moved outward from the boundaries of the central heaven. The Only Son, along with the Eternal Spirit call on God the Father to direct more potential energy via the Seven Master Spirits, into seven different areas inside the band, which was forming around the Central heaven.

One Of Seven Heavens Within The Heaven Of Heavens.

Because of the battle between gravity and friction, the spiritual energy contracted then exploded and expanded then because of gravity; it slows down again. That process happened over and over until slowly, more and more matter began to form. Spirit was first, then matter. In other words, the matter is created from spirit. Both originally came from the potential spiritual energy from God. It's just vibrating at a different frequency. When it reached its appropriate size and force, a triune of Ancient of Days were assigned. And they assumed their positions of authority. As the energy within and under their authority formed into a nebula suitable to form a Galaxy a Creator Son of the order of Michael, a great prince, would be brought forward and given their kingdom. (Read Daniel 8: 9-14. You see it?)

Local Universes (A Galaxies)

Is reigned over by a Creator Son and his conjoint worker the Mother Spirit. She is an exact duplicate of the Eternal Spirit (often referred to as the helper). Her work in the physical is perceived as the Holy Spirit. A Creator son's primary function is to create, control, and administer his creation. While the Mother Spirit is to work with the new life creatures and create an inhabitable living temple (humanity) wherein God the Father may project a portion of His awesome power and experience all of His creation. That gives that new living creature an opportunity to experience God the Father. Via this great plan of creation, we are afforded a method in which we have a direct line of communications with the eternal creator. In fact, we are given the opportunity to be a part of that great creation. We are sons and daughters of God who have taken the gift of free will and are experiencing Gods creation. As sons and daughters of man, we were given authority over all of the earth. We can experience, learn, and ascend all the way up the chain of creation and become one with the One God.

Informational Data of Jesus, Kingdom

When the need arose, after the manifestation and establishment of the Holy Triune, the Isle of Paradise, and the central heaven, the Father and the Son called into being the Creator Sons of the order of Michael. Our creator son planned out his creation with the Paradise Architects of the Master Universe. Those plans were all approved. The Master Force Organizers projected forth the power and energy of God into the third heaven. That action created the potential power of the seven super universes. The Ancients of days came and assumed the authority of their prospective throne. (Daniel 8:9 & 10) As a vast number of nebula began to form, our creator son went before the Ancient of Days. He chose and was given, this Andronover Nebula as the place for his creation. (Daniel 8:13 & 14)

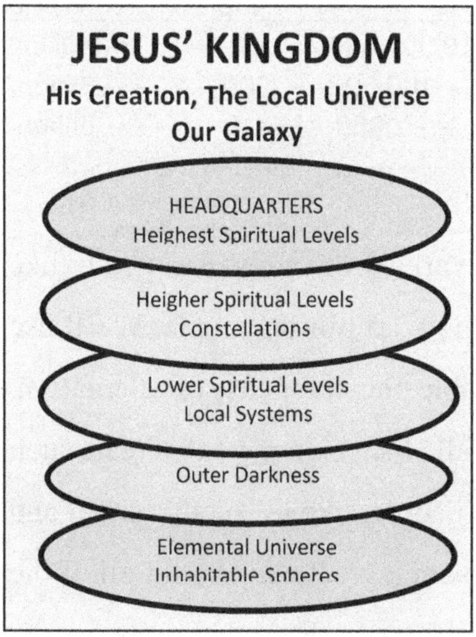

As the energy became responsive to local or linear gravity, the power directors assumed control and slowly brought the forces together sufficiently for the Creator Son to enter in and begin His long and tedious creation.

The exact center of our galaxy is the location of His headquarters. It was the first thing to be built. **Constellations** are the building blocks for a galaxy. There are 100 constellations in our galaxy. **Local systems** are the building blocks for constellations. There are 100 local systems in a constellation.

Solar systems are the building blocks for the Local Systems. They contain worlds which are the **Inhabitable spheres**. There will eventually be approximately 1,000 inhabitable worlds in each local system.

The creative plan is set up as a gradual evolvement and progressive development of the physical, intellectual, spiritual nature and capacity of the creatures inhabiting the various spheres

Numerical Organization And Population

Galaxy	Constel-Latations	Local Systems	Inhabitable Spheres	Aprox. Population Approx. ± 5 Billion People In Each World
Galaxy -1-	100	10,000	10,000,000	50 --- Quadrillion
Constellation	1	100	100,000	500 Billion
Local Systems		1	1,000	5 Billion
Inhabitat Worlds			1	1--Varies From a few to Like our world aprox. 7, Trillion People.

The number of inhabitants varies from a few early on to like our world, which is now approaching 7.5 billion people. We'll use an average of 5, billion per inhabitable planet for ease of calculation.

Anyone that thinks God in all His wisdom would create such a vast creation and only put living people, like us, on this small and insignificant planet we call earth is on a great ego trip for all of the humanity and is exercising self-aggrandizement unlike any humanity has ever entertained.

The basic unit for measuring things in our Galaxy is in light-years. That is how far light will travel in one year.

The Speed Of Light

Light travels 168,262 miles a second in a vacuum.

Light travels	186,000	miles a second.
Multiply that times	X 60	to get the distance for 1 minute.
Which equals	11,160,000	miles per minute.
Now multiply that times	X 60	to get the distance for one hour.
Which equals	669,600,000	miles per hour.
Multiply the answer times	X 24	to get the distance light will travel in one day.
Which equals	16,070,400,000	miles per day.
Multiply the answer times	X 365.25	to get the distance light will travel in one year.

That equals approx. 5,869,714,000,000 miles in one light year. (Approx. 6 trillion miles).

Our galaxy is approximately 110,000 light-years across and about 1/3 that in depth.

The distance, the amount of power, and energy contained in just one galaxy is immense.

Approximate Size Of Our Galaxy

IN LIGHT YEARS

	GALAXY	CONSTELLATION	LOCAL SYSTEM
WIDTH	**110,000 lt. yrs.**	**1,100 lt. yrs.**	**11 lt. yrs.**
DEPTH	**36,000 lt. yrs.**	**366 lt. yrs.**	**3.6 lt. yrs.**
VOLUM.	**3, billion 960, million**	**402,600 sq. lt. yrs.**	**39 sq. lt. yrs.**
CUBIC	**440, thousand cubic lt. yrs.**	**44,733 cubic lt. yrs.**	**4 cubic lt. yrs.**

All of this is approximate measurements given to impress upon you the magnitude and vastness of Jesus' kingdom. We will learn more about this, our great Local Universe, as we continue. But, perhaps we should shift our attention to the bottom of the preceding chart since we have built a foundation on which to climb up this ladder of spiritual understanding called the ascension process.

The Local System

Let us start where we are now on our planet. Our world is spiritually called Urantia, by some people. The prince of our world joined the rebellion with Lucifer. Because of his fall, we now know him, (the prince of this world), as the devil. Urantia is a part of our Solar system, which is spiritually called Monmatia, which is one of approximately 500 in our local system, which is spiritually called Satania. (This is pronounced like the material satin.) Its headquarters is the great spiritual city of Jerusalem often spoken of in the Bible. Satania will eventually have 1,000 inhabitable worlds. Satania is important to our orientation because it is where Lucifer was in a position of authority before his erring ways led to the rebellion or war in heaven. Satan was Lucifer's 1st. Lieutenant or liaison officer who traveled around spreading Lucifer's misguided teachings throughout the local system of Satania. Many people believe that Lucifer was an Archangel; he was a **son** of the morning (KJB), which designates him as being a SON, not an angel. His position of authority is the chief principality, or son, given authority over a local system. He had many spiritual personalities and angels in his command. If he, Lucifer, had bestowed himself as an archangel (which is a requirement for any spiritual son who is placed in authority over lower personalities, he must bestow or give himself to them). He would have that as a part of his titles, but he would still be a **son of the morning**. It appears that he is being confused with a morning star, which is the firstborn in a local universe. He is the chief executive of the creator son. There is only one Morning Star in each local universe or galaxy. Ours is Garble. A Morning Star is not a Son of the morning, nor is a Son of the Morning a Morning Star; these are two different positions of authority. In fact, it was Gabriel a Morning Star that defeated and removed Lucifer from authority. He, Gabriel, obtained permission from the Ancient of Days to confront Lucifer and retrieve all that would return to the ways of righteousness.

Orgazinatol chart of Christ's Kingdom

The Creator Son, Christ Michael

DIVINE MINISTERS The Mother Spirit Of Our Local Universe, duplicate of the Etermsl Spirit, and is the conjount worker with

THE HOLY SPIRIT IS THE MOTHER SPIRITS ACTION IN THE WORLD.

THE SEVEN MIND SPIRITS

Intuition, Wisdom, Knowledge, Understanding, Courage, Counsel, And worship these are the 7 columns or tools the Holy Spirit uses to create her building. Man is a builder too, and man is the building, a house wherein the Lord may dwell.

Angelic Host

MORNING STAR Only one In a Local Universe.!st. Born. Gaberiel is the Chief Administrator, Chief Executive, Chief Diciplinarian, the Chief of chiefs.

EVENING STARS Gabriel's assistants, teachers and assistants to principalities

ARCHANGELS assigned as needed throught the Local Universe. No Archangel of any order has left their post.

STARS ARE ANGELS

CHERIBUM

SERIPHIM

WATCHERS

GUARDIAN ANGELS

MANY OTHERS TO NUMEROUS TO MENTION.

Principalitie

MELCHEDEKS 2nd. Born High priestem emrgency counselors and advisoer.

There are 100 Constellations in a Galaxy. There are 3 Constellation fathers in each.

100 LCALA SYSTEMS in each Constellation. Each is reined by the Sovereign Son of the Morning, a brilliant ray of spiritual light. Lucifer was of this light. He was appointed over our Local System. He will later become the originator of the war in heaven.

SOLAR SYSTEMS with a Prince of that world over each inhabitable world. There is to be 1,000 in each Local System.

SONS AND DAUGHTERS OF MANKIND our world is inhabited by nearly 7 trillion people. We are the stewards of this world.

Powers

Spiritual entities to maintain and controll God's energy throught the Local Universe.

SATAN 1st. Lt. for Lucifer. It was he who brought Lucifer's eronious teachings to this world.

INFINITE Approx. 20 Approx. 18

<---------<------ --<------------------------Trillion------------------------------------Trillion

Years ago Years ago

The Beginning of the Time Line

Before anything was, there was this One force, One power, one essence that One is that Which we call:

THE ONE GOD

Before time, before space,Sometime in infinity A vast amount of Absolute Pure spirit was projected out From the One God and is Called Paradise.Paradise is the Powerhouse and the absolute center of creation.

The One God projected His total essence into Paradise and made it His permanent abode. We can know Him as God, the Father.

At Approx, 20, Trillion Years Ago Marked The Start Of Materialization Of Our Grand Universe: The Central Heaven And The Heaven Of Heavens.

Approx. 18, Trillion Years Ago Marked The Start Of Materialization Of Orvonton The 7TH Super Universe. In The Heaven Of Heavens.

The seven master spirits directed or funneled the awesome power from the One God into the firmament causing seven Super Universes to materialize. That spiritual power was directed by the mental energy of thoughts of God the Father and the Son. They were brought together as two different dimensions. Where they came together or overlapped the creative forces of the Eternal Spirit, through the Seven Master Spirits, brought the material mass into existence giving birth to the cosmos, our physical universe.

The process of bringing forth life in the lower dimensions began as time and space, spiritual and elemental was made ready for inhabitants. That process is still unfolding as infinity lays itself before God and man. see **GENESIS 1: 9-31, DAYS 4, 5, AND 6.**

When The Thrones Were Cast Down

Approx
1-Trillion-------|-----------|-------------|----------|-------------|-----------|------------|---------|
1-Billion 1,Tril,50 Bil. 1,Tril
Years ago years ago years BC

THE ANCIENTS OF DAYS ASSUMED THEIR POSITION OF AUTHORITY OVER THEIR SUPER UNIVERSE IN THE NAME OF THE ETERNAL TRIUNE. OFTEN THEY WILL SPEAK IN THE FIRST PERSON FOR GOD THE FATHER. Daniel 8:9.

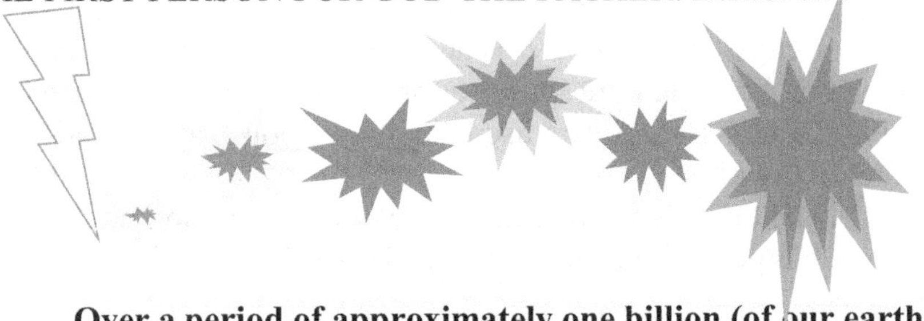

Over a period of approximately one billion (of our earth years)
the seven super universes began to materialize from the lower spirit and into matter.

Before the universes began to materialize and as the Master Force Organizers manipulated the space forces, by the plans of the Paradise Architects of the Master Universe, our great Prince Michael had worked out all His plans and had them approved by the Paradise Trinity.

This great Prince Michael, Daniel 8:13-14, 9:25, 11:21, 12:1, Rev. 5:4 & 6, is he whom we know as Jesus of Nazareth, Jesus the Christ, our Lord, and Savior. He was the creator Son of God, who is Creator, Lord, and Master of His kingdom; which is our local Universe, our Galaxy and includes our world. Jesus's kingdom is not of this world, but this world is of his kingdom. This Local Universe is his because he created it; it is ours because we live in it. He is our creator father. He is our lord and master. He is our big brother in the father. He is our best friend who loves us dearly. His greatest desire is to prepare his children for the great spiritual journey to the Eternal Father's house in Paradise.

The 7th. Heaven Our Super Universe

987 950 900
|Bill---------|-----------|----------|----------|Bill---------|--------------|-----------|------------Bill
Years ago Years ago Years ago

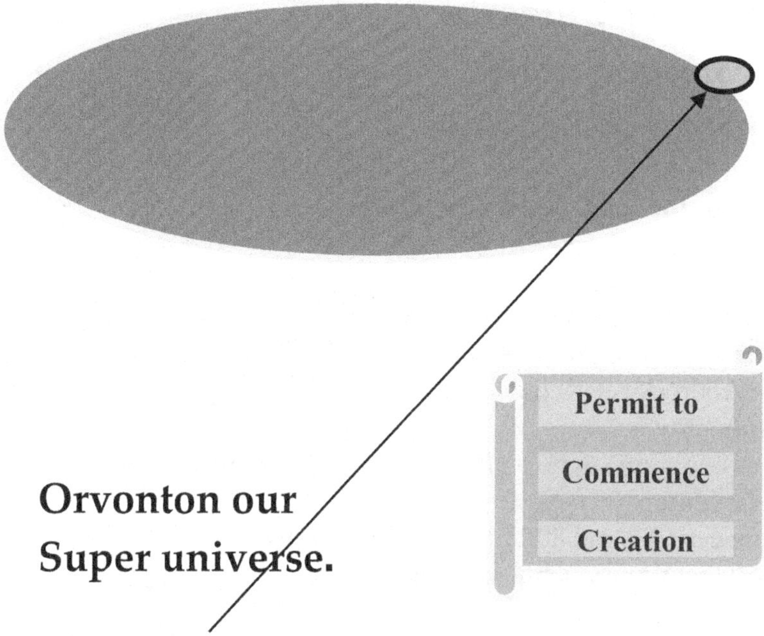

Space conditions in the then easterly segment of Orvonton materialization phenomena. Orvonton is one of the Super Universes.

Michael the Great Prince, The Mother Spirit and Gabriel the Archangel (the Morning Stars) were ready to begin the long and tedious creation of a new heaven and earth.

A permit was issued to the Super Universe Government to Dispatch force Organizers to get the space energy ready to start creation according to the mandates of the Ancient of Days.

900 850 800
Bill------------|------------|------------|---------|Bill---------|----------|------------|-------------Bill
Years ago Years ago Years ago

Orvonton is the 7th Super Universe Or heaven in the Heaven of heavens.

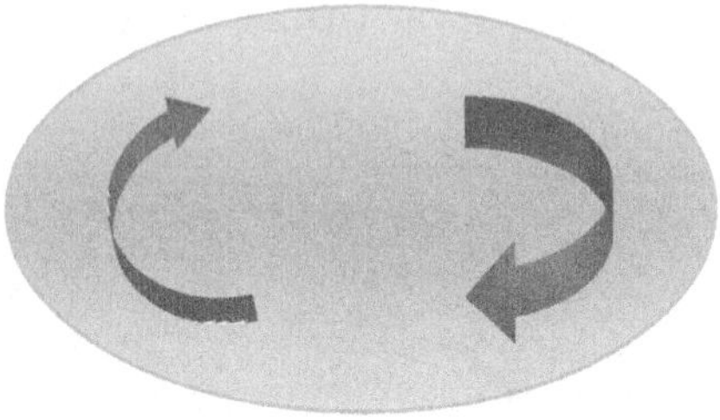

Within the Andronover Nebula
Which will become Our galaxy.
Force organizers
departs at right angle causing a swirling motion to take place in the Andronover Nebula gases.

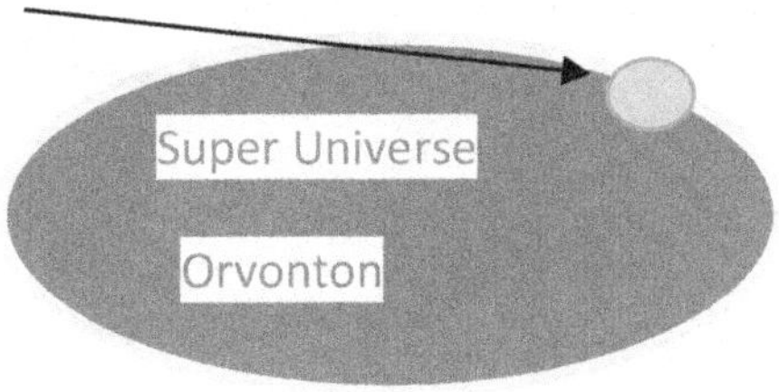

Andronover
Was beginning to be well organized as a magnificent primary nebula in Orvonton.
Between 900 and 800 billion years ago we see The primary nebular stage

GENESIS 1: 1 for our Local Universe, our Galaxy.

The early stage of creation is in a gaseous stage. It is circular in shape like a flattened sphere. It will gradually become spiral. As gravity begins to work on the energy of the gasses, they will contract, condense, and speed up. The energy will condense and tighten until it erupts. Every time, gravity will win out and pull the forces back again and again. It appears as though there is a constant struggle between these forces, but it is the natural process of creation directed and controlled by our Creator Son and power directors according to his plans. Those gasses are slowly but surely being changed to a material matter.

With the Primary Nebulae and the Mother Mass

800 700 600
Bill----------|-----------|------------|----------|Bill----------|-----------|-----------|-----------Bill|
Years ago Years ago Years ago

The early stages of our Galaxy.

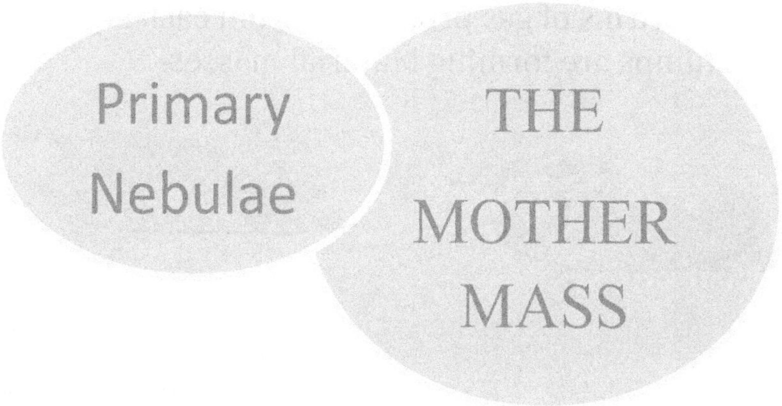

The Andronover energy mobilized into a circular gas cloud.At the height of the energy mobilization period and its maximum mass is reached. About 600 billion years ago

Between 800 and 600 billion years ago, The primary nebular stage continued. Gasses were changing into the physical matter because of the struggle between the energy of the gasses and as the constant growing gravity pulled more and more matter into existence. Space materialization was forming. Additional controllers were dispatched to afford support and co-operation to the power centers of this new material system. It would be like Matter was being formed from dark matter (or spirit-matter). As it reached its maximum size, the material began to whirl faster and faster.

The Secondary Nebular Stage

600 550 500
Bill----------|-----------|------------|-----------|Bill----------|------------|-----------|-----------Bill
Years ago Years ago Years ago

Arms of gas project out from each side.
Clumps are forming Material masses.

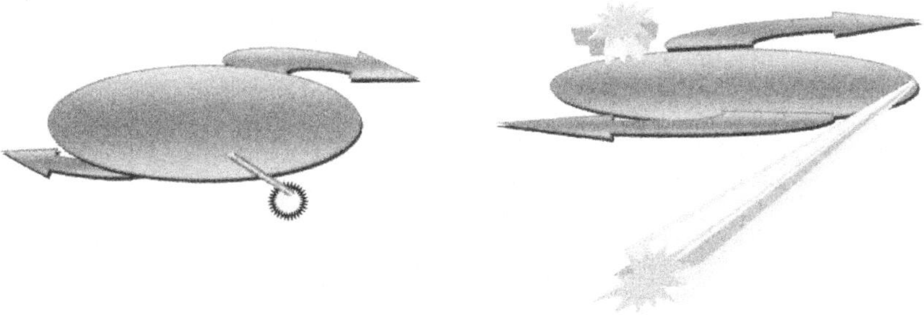

The first sun was born. This blazing streak broke away from the mother gravity grasp and tore out into space. Approximately 500, billion years ago. More newborn suns began to break away.

The enormous nebula began to gradually assume the spiral form, which is the nebulae history of most nebulae before they begin to throw off suns and stars. The rapid revolution of the enormous central mass and the weakening of the gravity control imparted caused the gasses to escape at both ends. As the arms extended the cooling and subsequent condensation caused knotting of systems and subsystems of physical matter whirling through space. As the nebula began to contract, the matter within began to swirl faster and faster. It was loosening the hold of the gravity force.

500 billion years ago, the first sun escaped from the grasp of the mother mass and began its venture outward into space many other Suns reached escape velocity. Some suns were recaptured while others continued in the evolutionary process of suns and planetary bodies.

At about 400 billion years ago, there was inaugurated the terminal phase of nebular condensation. That is the period which always precedes the final segregation of these immense space aggregations of energy and matter.

The Mother Mass Of Nebadon

400 **350** **300**
Bill------------|---------|-------------|-----------|Bill----------|------------|-----------|-----------Bills
years ago Years ago Years ago

Our local universe is reducing in size. Enormous heat from contraction and condensation of the central cluster is occurring.

Creation was taking place in both dimensions: spiritual and the elemental universe. As spiritual energy (dark energy) expanded out and condensed, spiritual matter (dark matter) began to form. Spiritual entities began to create the headquarters and spiritual worlds. Along with the spiritual material worlds. Together they are called SPIRITUAL Dimensions. The physical worlds of the elemental Local Universes were being formulated from the gases within the galaxy.

NEBADON: is the name of our Galaxy which is our local Universe.

SALVINGTON: Is the headquarters Of our Local Universe, our galaxy.

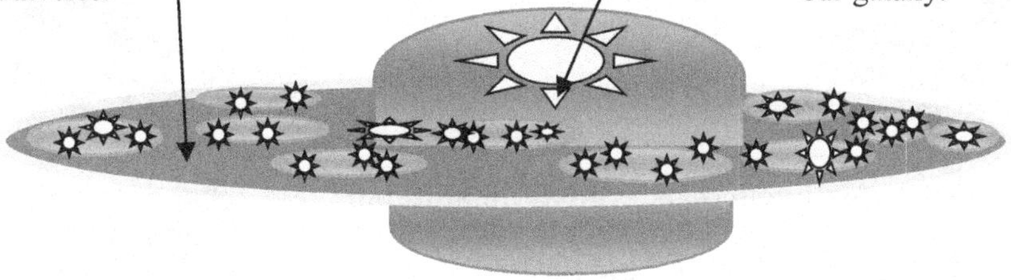

Represents Constellations: They are built first spiritual with a spiritual headquarter and then physical.
Each constellation will eventually have 100 local systems.

Represents Local Systems: They too are built first spiritually with spiritual headquarters and then physical. Each Local System will eventually have a variety of solar systems with a total of 1,000 inhabitable spheres.

It was scarcely a million years after this epoch that Christ Michael of Nebadon, our Creator Son of Paradise, selected this disintegrating nebula as the site of His adventure in universe building. Almost immediately the architectural worlds of Salvington and the one hundred Constellation headquarters groups of planets were begun. It required almost one million years to complete these clusters of specially created worlds. The Local System Headquarters Planets were constructed over a period extending from that time to about five billion years ago.

The Great Prince Michael Arrives On Salvington

300 **250** **200**
Bill-----------|---------------|---------------|Bill--------|---------|----------|-----------|--------Bill
Years ago Years ago Years ago

Michael's creation is registered as inhabitable

for progressive mortal ascension.

300 Billion years ago Andronover nebula solar circuits were well established, and the nebular system was passing through transit of relative physical stability. About this time, the staff of Michael arrived on Salvington, the capital of our local universe, the government of Orvongton, our super universe, extended physical recognition to our local universe Nebadon.

At about 200 billion years BC, some of the newborn suns cooled enough that some of the planets were suitable for life. They are the oldest inhabited planets in our local universe.

Complete universe mechanism of Nebadon, our local universe, began to function

.

200 **150** **100**
Bill-------|-------|---------|------------------ |Bill------|----------|------------|---------------Bill
Years ago Years ago Years ago

Two Phases Of Creation Were Simultaneously Being Created.

THE **SPIRITUAL** DIMENSION WAS CREATED. FIRST THE HEADQUARTERS, SALVINGTON, AND THEN THE 100 HEAQUARTERS OF THE CONSTELLATIONS.

THE **ELEMENTAL** UNIVERSE WAS BEGINNING TO MATERIALIZE FROM THE GASSES

GENESIS 1: 5 the 1st day or stage of creation for our Local Universe, our Galaxy.

Construction of the headquarters of Salvington began about 400,billion years ago and took about 1 billion years to physically build. It was not physical as of the elemental world we know.It would appear just as physical to you as the elemental world is to us physicalcreatures.

At about 15 billion years ago, our System Sovereign, Lucifer, Son of the Morning was chosen by Michael. Lucifer was a Lanonandek Son of Nebadon; he was a Brilliant spiritual son created by Michael.Lucifer was 37 of 700,000. He was not an ascending son. He was a brilliant ray of light, a great administrator and leader.

During this time we see Nebadon, our local universe was constructed with 10 circuits of 49 worlds each.

Melchizedeks Second-born of Michael and the Mother Spirit. High Priest of the Most High God: 1st. Assistant with Gabriel the Morning Star.

Vorondadeks: Born of Michael and the Mother Spirit. They may occupy a variety of positions with high authority. They are a little more physical.

Lanondeks: Most versatile of Local Universe administrators also born of Michael and the Mother Spirit. Lanondek sons are more physical than Vorondadek Sons.

Life Carriers: Plan and initiate life, catalyze lifeless material with the spark of life. Life Carriers are created by the creator son and the Mother Spirit.

Finalizers: Personalities that have ascended from the lower physical dimensions up the ladder to paradise and have found the Father.

Brilliant Evening Stars: Archangels of the highest. Assistants to Gabriel, and report to Him in his absence.

Archangels: Highest order of angels that may be assigned anywhere under Gabriel, or to one of the administrators in the Local Universe.

Spirit-fused mortals: Physical personalities that are attuned to and respond to the indwelling Spirit of God. **Seraphim**: Members of the angelic corp. May be assigned to an individual guardian, or groups; Great teachers of truth; and have a large list of possible tasks to which they can be assigned. More than 500 million are assigned on Urantia, our planet

.

Organizational Chart of Lucifers Domain

100 **75** **50**

Bill------------|--------------|---------------|--------------|Bill---------|-------------|------------|------|------Bill

Years ago Years ago Years ago

This part of the chain of command was not completely established at this point; it is included here so you can see how it relates to our planet.

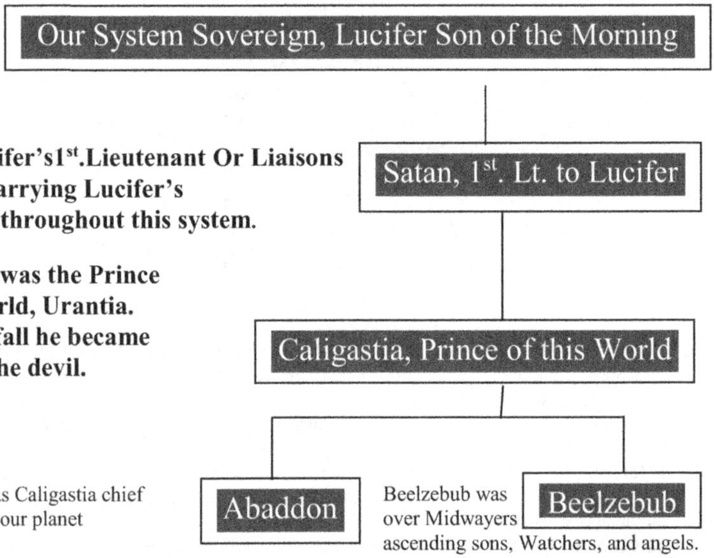

Satan **Lucifer's1st.Lieutenant Or Liaisons Officer Carrying Lucifer's Messages throughout this system.**

Caligastia **was the Prince of this world, Urantia. After his fall he became know as the devil.**

Abaddon was Caligastia chief of Staff over our planet

Beelzebub was over Midwayers ascending sons, Watchers, and angels.

At about 100 billion years ago, Lucifer picked and assigned Satan, who was anxious to be in authority as his liaisons officer and administrator. It was Satan who took Lucifer's word throughout our Local System. He would visit the many princes of the developing worlds that supported life. Thus, he traveled extensively. Later he will be the first to accept Lucifer's erroneous teachings and proved to be a great spreader of that teaching. But for now, he and Lucifer performed their task brilliantly. For those of the spiritual world, it was a great blessing to serve under such brilliant Sons of God like Lucifer and Satan.

GENESIS 1: 10-13 BEGAN TO BE FULFILLED AS WORLDS BEGAN TO DEVELOP. GENESIS 1: 14-19 were being fulfilled FOR EACH WORLD.

The First Step Of The Second Nebular Stage

50 **25** **10**
Bill------------|--------------|---------------|--------------|Bill---------|-------------|------------|------|------Bill
Years ago Years ago Years ago

The 1st. step of the 2nd. Nebular stage was coming to an end. This step of sun dispersion gave birth to 876,926 suns.

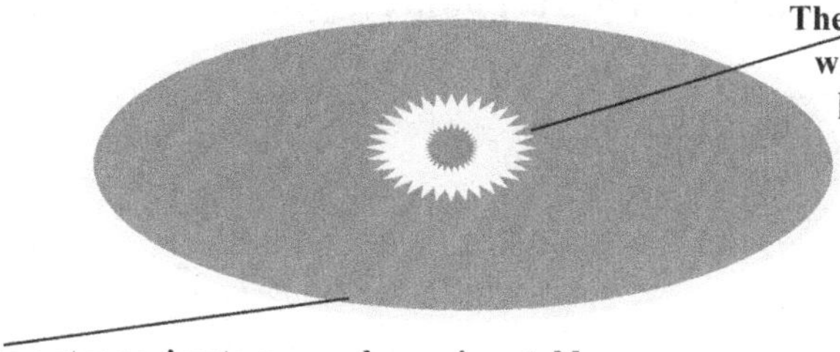

At 10 billion years ago terminal Eruption began. Only the outer regions were safe.

The outer perimeters were becoming stable, but the central nucleus was coming under more and more pressure and heat. The gravity-heat struggle becomes immense, but the heat will win the battle and will cause a nuclear explosion that will start the second phase of sun dispersion.

The end of the great Andronover Nebula

10 9 8 7 6 5 4 3 2 1
Bill---------|-------------|------------|------------|-----------|-------------|----------|----------|-Bill
Years ago Years ago Years ago

The great ANDRONOVER NEBULA is no more.
A two-arm spiral barrel galaxy. (Similar to ours.)

GENESIS 1: 6-8 for our local universe, our Galaxy.

THE BIRTH OF NEBADON. Terminal eruption begins.

OUR LOCAL UNIVERSE. Was being born. It was at the height of the terminal eruption of lager terminal suns.

DURING THIS POINT OUR SUN IS BORN

This final eruption gave birth to 163,702 suns most of them were solitary orbs. Our sun was the 56th from the last. The total number of suns and sun systems having an origin in the Andronover nebula was 1,013,628. The number of our sun solar system is 1,013,572.

At this point the center still burns with a reddish glow giving off heat and light to its remnant planetary family of 165 worlds.

Christ Michael's headquarters, Salvington, and the Constellation Headquarters have long been in full operation. About 5 billion years ago, almost all of the Local Systems were completed spiritually and began to be fully operational.

The Birth Of Our Solar System

5 **4** **3** **2 Bill**
Bill---------|-----------|------------Bill|--------------|------------Bill------|----------|-------700Mil
Years ago Years ago Years ago Years ago

Our sun was a comparatively isolated sun formed by nearby matter from the recent upheaval of its birth.

4 Bil. 500mil. The Angona System passes near our Sun. The dark giant in its center; was solid, highly charged mass possessing a tremendous gravity pull. This system did not steal any of the Energy from the sun, but it did pull a long arm Out from the sun that will develop into our solar system.

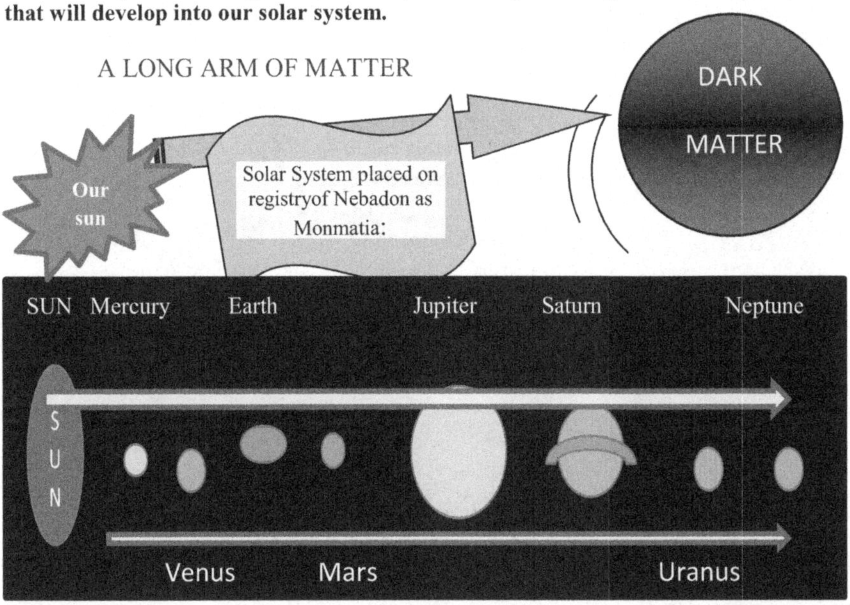

The energy that was pulled out from the sun will remain detached from the gravity pull of the sun. The outer sectors began to form independent bodies of matter and immediately went into orbit around the sun. That great column of gas began to form as asteroids and meteors began to be drawn into larger bodies that eventually became our solar system. Most of the planets were first formed as miniature planets that would grow as gravity pulled more and more matter to them. While still in gasses/matter stage gravity pulled the heavier metals like iron into the inner core. About 1 billion years ago marked the beginning of our planet.

Development Of Urantia Our World

2 Bill **2Bill** **2 Bill**
700 MIL.--------|--------|----------|--------|---------|500 Mil---------|----------|---------|----------|
Years ago Years ago Years ago

The Meteoric Era, the Volcanic Age, and the Primitive Planetary Atmosphere

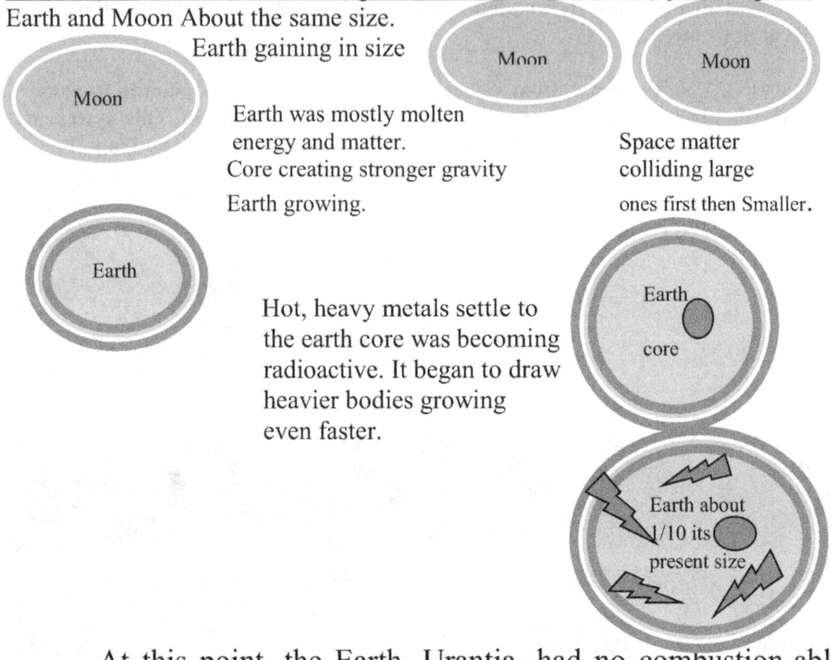

At this point, the Earth, Urantia, had no combustion-able atmosphere that would burn up the incoming matter. Because of the stronger gravitational pull from the core, collection of space matter caused the earth to increase in size, and therefore, it outgrew the moon. The surface of the earth was more molten energy and matter; it was easier for the heavier metals to sink into the core. The solar system was functioning much as it is today. Volcanic activity dates from about this time. Radioactive studies show the earth to be about 1 billion years old, but it does not take into account that these radioactive materials did not appear on the earth until later.

2 Bill **1Bill** **1 Bill**

|---|700---500 Mil

Years ago Years ago Years ago

LUCIFER was beginning to harbor some feelings of self-aggrandizement. His self-contemplation will prove to be most disastrous. His rebellion started as pride of self, which led to self-deception. He convinced himself that his rebellion was good for the system. This process of self-deception was a slow process building up in his mind while professing loyalty to the creator son. Lucifer served as he was supposed to. It would be sometime later that these seeds of selfishness will lead to an open rebellion against Michael, the heavenly host in authority, and he would deny the existence of the One God.

1 Billion 500 million. Years ago.The earth is about 2/3 it's present size.

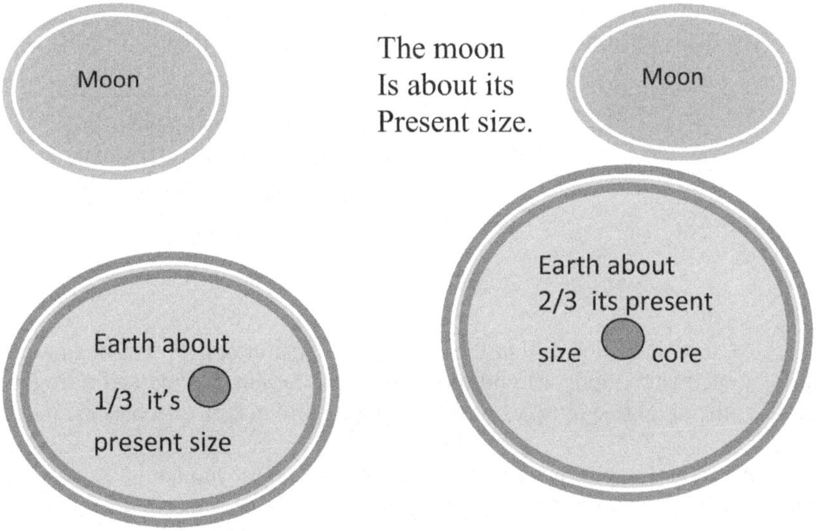

At 1 Billion 500 million years ago, volcanic action was at its height. The whole world was virtually a firing furnace; this is the volcanic age. Nevertheless, crust consisting virtually of the comparatively lighter granite is gradually forming.

A primitive planetary atmosphere of the volcanic age offered little protection against collision impact of meteoric swarms. Millions upon millions impacted the earth's surface.

Early Earth Developing

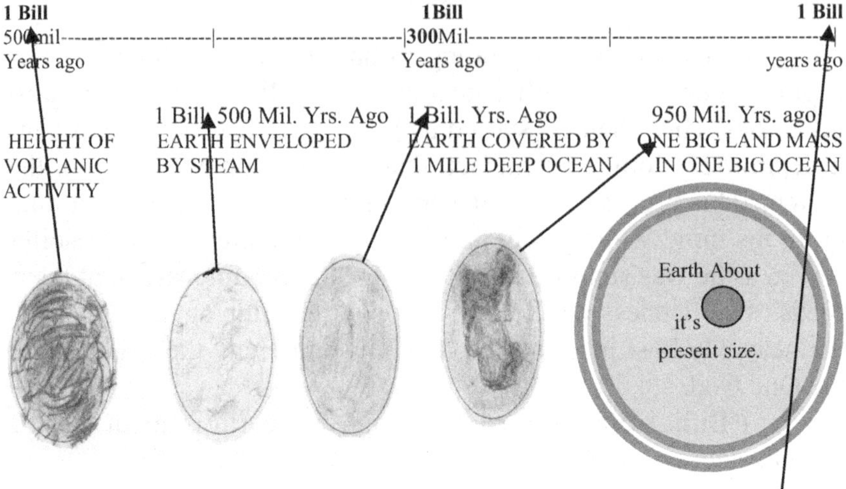

THE EARTH WAS BEGINNING TO DEVELOP AN ATMOSPHERE.GENESIS 1: 9. FOR OUR WORLD.

The earth is placed on the physical registry of NEBADON and given the name URANTIA. 1 Billion years ago.

At about 1 billion 400 million years ago, an atmosphere was beginning to form; some water vapor, carbon monoxide, and carbon dioxide and nitrogen chloride. Little or no free nitrogen or free oxygen; burning meteors consume most of it.

As water formed in the atmosphere, it fell into the molten lava, causing steam to rise. That steam completely enveloped the earth. As the rain fell, it cooled the earth's crust, and eventually, the steam condensed and disappeared, creating an ocean that was about 1 mile deep and covered the entire planet.

The real history of Urantia began with the cooling of the earth's crust sufficiently to cause the formation of the first ocean. Water vapor condensed and cooled the surface. The water was not salty; it was primarily freshwater with the chlorine combined with various metals causing the water to be slightly acid. Tides were coming into play.

A single landmass was beginning to appear as the crust thickened and volcanoes erupted.

Urantia Was Registered

on the major and minor sectors of Nebadon.

1 Bill 900 Mil 800 Mil 750 mil yrs. Ago 1st crack. 700 mil 600 Mil

|----------------|------------------------|--------------------------------------|------------------------------------ |

Years ago Years ago years ago

Christ Michael's **1st** bestowal as a Melchizedek, High Priest. He was accepted as a chief Melchizedek.

1st. a scouting party arrives on Urantia (our world).Life carriers were permitted to institute new patterns of mechanical, chemical and electrical mobilization of life patterns.

Christ Michael's **2nd** bestowal as a Primary Lanonandek son. He was placed in authority where trouble was brewing. He was admired as being righteous and just in all ways.

Christ Michael's **3rd** bestowal as a Material son, a planetary Prince. On an isolated quarantined planet. He had to work alone. He affected the repentance of their prince, his staff and restored them all to the Paradise Father.

On Christ Michael's **4th** bestowal he was secretly assigned to the Seraphic Teaching counselors, private secretary to 26 master teachers. And he worked secretively on all levels of angelic service.

Tectonic 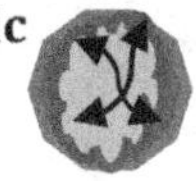 **plates Moving.**

World 606, Urantia, our world, was assigned to Satana our Local System.

. The Great Prince Michael served on 4 emergency missions and was completely accepted as one of their order. He then returned and assumed his position of . authority. When he left, only a few knew of his mission. This was the procedure for 6 of the 7 bestowals. It would be announced that he would be absent and authority was turned over to Emanuel, his older brother., It was announced that Urantia was going to be the location of his 7th. And final bestowal We of this world will get to know our Beloved creator King as Jesus of Nazaret.

THE BESTOWAL PROCESS: God's spiritual children have eternal life. They are given authority over lower creatures and agree too, bestow themselves (give themselves.) to those lower creatures on each level in their authority. Thus,they are given an opportunity to know and learn, experientially, and gain knowledge of their creation. They are accepted as one of that order. His creation will have an opportunity to know their creator.That bestowal is an inherent quality of the Paradise Father and Son. This process is experienced through the Seven Master Spirits and Allows these great Creator Sons an opportunity to become one with each of the Eternal Triune It becomes their heart's desire to become one with the Father completely. Thus when we see one of them, we see all of them.

The Life Dawning Era

600 Mil	**500 Mil**	**400 Mil**
Years ago	Years ago	years ago

Earthquakes continue,upon the earth mountains increase as crust thickens. 1/3 earth is land mass. 1st continental cracks appear Creating finger- like lakes of brackish saline waters. No moisture; no ice yet.

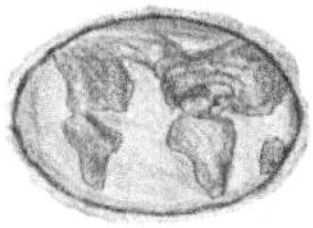

The Americas Move west. Africa moves south.

The earth develops a Dense core core

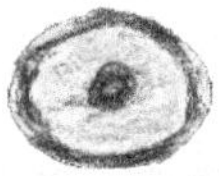

up to 24,000 tons per sq. inch.

The life dawning era saw a lot of things beginning to happen on Urantia. At about 600 million years ago the life carriers surveyed the planet and decided to wait for more continental separation and the water to become a little more saline. They also wanted favorable conditions in the atmosphere. At 550 million they implanted the Marine life seeds

Life Carrier corps initiate identical life patterns of marine life implantations in three locations.

1. Central or Euro-African.
2. Eastern or Australa Asian.
3. Western embracing Greenland and the Americas.

The Continent drift carried life into all parts of the world.

With the Non-crystallized crushing sea of molten basalt.

A little over 500 million years ago Vegetation life begins and spreads.

Around 450 million years ago Transition from vegetation to animal starts.

Continents and ocean bed floating on molten base.

God's Life Carriers can, when permitted, bring the original life forces into a planet or they can create it from the existing material life forces that exist on that planet. The latter was chosen for Urantia.

The Flood Period

Genesis 1: -20-23

400 Mil |---|------------------------------------|--------------------------| **300 Mil** **200 Mil**

Years ago Years ago years ago

The earth crust and plate tectonics Move not only sideways, but Up and down as well.

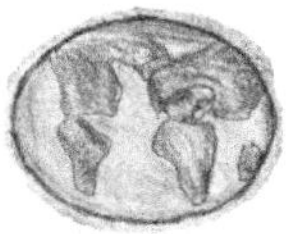

Marine life slowly dawns.

Marine life consisted of: seaweed, one- cell organisms, simple sponge, trilobites, and coruscations, shrimp, crabs, and lobsters: Trio-bites were dominating living creatures at about 400 mil. Years ago.

During this period saw the rise and fall of land masses. The Americas, Europe, and Africa (Except Asia) sink below the water over and over, again and again and, again over millions of years.

As land life increases.fish began to appear. Carbon dioxide decreasing in freshand. saline water.

Sharks become predominant

Marine life moves on land.

Frog age begins

Marine life spreading across the land when underwater. Fossils deposited worldwide. Vertebrate and scale were appearing. No land animals except some worms.

At about 300 Million years ago .The 1st great flood starts

At about 200 Mil. Years ago Last and least extensive sinking of land occured.

During this period most plant life near water's edge: fern family moves inland covering most of the land. Bacteria kill most small plants.

Coal is being laid down world wide.

Land is over-run by ferns.

Fern age

400 Mil

Rise & fall of land mass

Land going down again
Land up again.
300 Mil 200 Mil
Land under Water 10, to 15 thousand feet

Christ Michael's **5th** bestowal. Having finished the 4 higher levels of spirit life he moves down into the lower spirit realms of ascending mortal origin. He was tried and tested in all ways. He was becoming a part of the mortal personalities.

There were no air breathing creatures because of too much carbon dioxide in the air.

No air breathing creatures but Later Insects start to appear

Trees without leaves appear. Foliage develops later.
OCEAN CLIMATE MILD AND UNIFORM.

200 Mil |------------| **150 Mil** |------------| **100 Mil**

Years ago — Years ago — years ago

LUCIFER

He was showing some evidence of discontent and starting to show his self-aggrandizement, but he still operated and functioned as a loyal servant of Prince Michael. Some of his loyal associates would occasionally bring it to his attention. He would pretend to hear but more and more he turned a deaf ear to them.

Involution; was occuring some of the spiritual sons of God were beginning to get involved with creation before it was made ready.

Christ Michaels **6th.** Bestowal as a Morontia mortal of evolutionary ascension. He was tested and proven magnificent in all ways. This was probably his most wonderful bestowal. He was about doing the Father's will.

Through his bestowals, this paradise son was lovingly learning his created creatures. They in turn, were learning to love Him, their creator.

1,800,500 years ago.

Fresh water And salt water Lakes.

JORASSIC AGE lasted 125 Mill. Years

Rich marine life Mammals, appear but Soon disappear.

REPTILE DINOSAURS BECOME MONARCHS.

DINOSAURS Dying out.

LANDMASS CONTINUES TO RISE AND FALL, BUT IT DOES NOT COMPLETELY SUBMERGE, AND LAND MASS GAINS IN SIZE.

200 Mi — **150 Mil** — **100 Mil**

Mountains on the rise causing drastic changes in wind and weather, which kills most life

New marine and land life develops stronger seed and begin to flourish and replenish the ocean and

Genesis 1 24-25 Being fullfiled.

Reptile age drawing to a close.

Egg laying reptiles appear and multiply rapidly: Crocodiles, scaled reptiles, sea serpents, carnivorous, small reptiles, later small brain, big body (1lb. brain with 40- ton body).

Most large creatures, land and sea, dying out from lack of brain size to cope.

Now land animals and plants become predominant.

For the most part, the water mothered and nurtured life up to this point.

100 Mil |------------| **50 Mil** |------------|------------|

Years ago Years ago years ago

CRET ACEOUS PERIOD

ATLANTIC AND INDIAN OCEAN MUCH AS IT IS TODAY.

End of Westward & Southern continental drift. At 75, mil. Land mass moving up and down in areas, mostly up with mountain peaks rising.

Flowering plants.began to appear.

Bird life on land.

Fig trees, Magnolias, tulip trees, & palms.

1st. Mammals
Carnivorous, small springing type dinosaur.

BIRDS 3rd type flying creature, 1st. true bird. 55, mil.

At 95, mil. Years ago Great Volcanic activity Raising land mass. Continued elevation thrusting upward. Bering Strait closed 85, mil. One of largest lava flows 65, mil.

Artic regions still enjoying the warm climate at that point.

At about 50, million years ago the 2nd great continental Submergence.

Rising land close to present day elevation.

Seed bearing plants and trees, the 2nd most important step in evolution, Weeds spread, floral developing. Great plant life of evolution.

FIRST ADVENT OF THE ICE AGES: 2, mil. 1ST American Glacier. 1.5 mil. Retreats.

Huge elephants, horses, deer, oxen, camels, rhinoceroses, the cat family, dogs etc. Approx. 3, mil years ago.

Urantia, OUR PLANET REGISTERED as inhabitable. Approx.1 Mil. B.C.

First primordial man evolved. Genesis 1:26-28. They were called Sons and Daughters of man.(1.5 Mi. - 991 Thou. Years ago).

Primitive Mankind Appears On The Earth

1 Mil **900, Tho.**

|---------------------------------|-----------------------------------|--------------------------------

Years ago years ago

At about 1,500, BC the PRIMATES, the 3rd. vital mutation suddenly appeared.

--.--

The Mother Spirit assigned The Seven Mind Spirits:

1. Intuition
2. Understanding
3. Courage
4. Knowledge
5. Counsel
6. Worship
7. Wisdom

They work within the minds of developing animals and mammals wherein the life carriers planted the life plasma.

Our world was given WILL DIGNITY:

Which means mankind had willpower, the ability to choose and worship and ascend.

THE BIRTH OF THE 1ST. TWO HUMAN BEINGS. (991,485 BC)Mankind began to feel he was more than just an animal. He had big toes and functional thumbs and walked upright. The descendants of these two are those referred to in the bible as the sons and daughters of man. The Indwelling spirit of God (A portion of God the Father that indwells in the mind of will creatures, like a mustard seed) began to influence mankind's thoughts.

FEAR joined with ignorance of natural phenomena gave birth to primitive religion.

At 550, mil, the life carriers implanted life plasma on this world it was now bringing forth life that has willpower, with the ability to choose.

--..--

Of the three implanting of life plasmas two began to show results, the western and the central. There were slight changes in the animal genetics that brought about great mutations and a new species; mankind: Lumerians were a humanoid species, preceding mammals, mid-mammals, to the primate with less hair, walking feet, working hands and a reasoning willful mind. They were beginning to respond to the mind spirits of God.

--..--

Life carriers prepared to leave this world for their work was done.

PRIMITIVE MANKIND developed the ability to think and make decisions for themselves; they decided to leave their ancestors. As they did they began to develop will dignity. These primitive people were more human and less animal and were called called ANDONITES. They lived in caves, discovered flint, fire and used skins of animals. They followed the movement of ice and were afraid to go back to the world of their ancestors. Cold weather forced them to be resourceful. Their knowledge grew from their experience. As primitive language began to develop they began to seek counsel among themselves.
FEAR WAS THE BEST PRESERVER OF LIFE

Start of the 3rd GLACIAL Somewhere along the line mankind sun in awe and fear and tried to understand them and their relationship with them

1 Mil **900, Tho.**

|--------------------------|-----------------------------|------------------------

Years ago years ago

900, Tho. **500, Tho.**

|--------------------------|------------------------------|--------------------------|------------------------|

Years ago . years ago

Early mankind became good hunters and providers for their families, the clans which became tribes. The best hunters would get their choice of women for sex. Their desire for sex greatly improved their hunting skills.

They were doing precisely what Genesis 1: 28 say's to do. "Be fruitful and multiply and subdue the earth."

These tribes would split up and move on to new land. Fighting and wars for good hunting grounds and sex began to break out.

A mixture of the variety of tribes gave rise to the NEANDERTHAL, spreading from the highlands to India, France, and China and down into North Africa. (850,-350, B C)

ONAGAR

THE 1ST TRUTH TEACHER

Brought peace to the various tribes by leading them in the worship of the "Breath Giver to man and animals."

Food was all-important to the tribes; he taught them their 1st prayer.

"O Breath of Life, give us this day our daily food, deliver us from the curse of the ice, save us from our forest enemies, and with mercy receive us into the Great Beyond."

Development of Superstitions and Sacrifices

They didn't understand the natural forces behind the things they feared and developed superstitions which somewhat resembled religion. They had a deadly fear of clouds, more especially of mist and fog. The sun was good because it returned every day, but they could hardly be called sun worshipers, they lived rather in fear of the darkness. Everything was fine when the moon was bright, but when it grew dark, they grew panicky and began to sacrifice some of their most able men and women to appease the moon. To them, it appeared that their sacrifices worked because the moon would return to a brighter stage. Remnants of some of the superstitions and fears of the unknown, and the need to give sacrifices to the unknown (or gods), still lingers till today.

GAME WAS ABUNDANT

Many species of deer, as well as elephants and hippopotamuses, roamed over Europe. Cattle were plentiful and horses and wolves were everywhere.

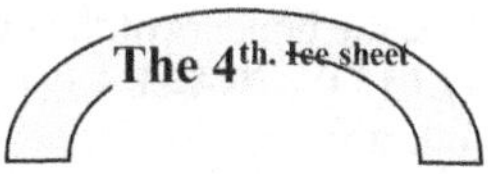

At 750 Tho. Years ago The 4th ice sheet was well underway receded about 400 tho.

THE GLACIERS MOVE PEOPLE AND ANIMLS

The glaciers, again and again, pushed man and animals back and forth across the land. These Neanderthal races had little or no religious feelings. Their primary motivation was fear and more fear. Their primary drive was for food, shelter, and sex. Their progress was slowed down and was beginning to regress with time. Tribal fighting and wars destroyed some tribes and brought many to near extinction.

The development of this early mankind was different from the normal development of mankind in that it superseded the colors or six races of people, about 500 thousand years ago there occurred, in one generation, one family, and offspring of different colors. Their skin changed as it was exposed to the sun, and their offspring was born with the same pigmentation coloring.

500. Tho. **400, Tho.**

|---------------------------|------------------------|---------------------|------------------------|

Years ago years ago

The advent of the 5th ice sheet.

LUCIFER
Appoints and assigns Caligastia as Prince of this world.

THE PRINCE OF THE WORLD

Caligastia was of the second order of creator sons of God. He too was a brilliant personality and performed his tasks admirably.

PRINCE OF THE WORLD
ARRIVAL AND MISSION

The prince of the world arrived on this world 500, thousand years after will dignity was bestowed upon mankind. There were almost one-half billion primitive human beings on the earth scattered over Europe, Asia, and Africa. He and his staff established the headquarters in Mesopotamia which was about the center of the world population. There were a large number of angelic co-operators and a host of celestial beings assigned to advance the interest and promote the welfare of the human race.

Somewhere along the line, humanity began to worship the sun, the moon, and the stars.

THE COLORS OF NEW MANKIND

RED RACE: First to develop tribal civilization, developed bow and arrow. Waring among themselves weakened them so much the yellow man drove them out of Asia. They then migrated across the straights into the Americas.
ORANGE RACE: Builders of anything, everything, they would competitively stack stones just to see who could build the highest stack. Some of their ancestors were the builders of the pyramids.
YELLOW RACE: Established settled communities, agriculture socially and collectively.
GREEN RACE: Were less able and weakened by extensive migration. They divided into three tribes but were soon defeated by the orange race.
BLUE RACE: Invented spears, developed rudiments of many arts. They had the brain power of the red race associated with soul and sentiment of the yellow race. They would be the preferred race by the Adamic descendants later on.
So-called **WHITE RACE**; is descendant of the blue race mixed with the yellow race who later assimilated most of the violet race which was a mixture of all colors. These mixtures served to uplift mankind's genetic chain.
INDIGO RACE: Now called the black race was least progressive and migrated into Africa and remained there unless forcibly removed. They believed in "God of Gods", but soon forgot and maintained primitive forms of worship.

400. Tho. **300, Tho.**

|-------------------------|-------------------------|-------------------------|-------------------------|

Years ago years ago

The 5th ice sheet recedes.

Humanity began to take note of the Movements of the planets (Wanderers) and the different Positions the sun rose in the morning. They discovered it would reach a low point and start back in the opposite direction again (December 25th).

The shortcomings of the Prince of the world.

The prince of the world was slowly falling insidiously in love with himself, succumbing to the subtleties of personality pride. Later he will betray his sacred trust of planetary custody and would so horribly stain the fair name of his exalted order of universe sonship.

THE DEVIL

The prince of the world was among the most beautiful creatures. That beauty will be stripped of him and his true ugly grotesque personality will become known as the devil when he joins the Lucifer rebellion and he and some of his staff begin to assert their own authority and began the rape of the human species he was supposed to be protecting.

LUCIFER

Was a created spiritual Son of the morning called into existence by Prince Michael and was assigned as chief executive, Sovereign of our Local System. He was over 607 inhabitable worlds. The contemplation of self: self-aggrandizement, self-gratification, self-glorification, self-pride and egocentric selfishness developed into self-deception. He was becoming more affirmed in his beliefs. Lucifer was becoming unhappy with the universe order. He developed Lucifer's declaration of liberty.

SATAN

Was Lucifer's 1st assistant or, 1st Lieutenant. He at first disagreed with Lucifer, but soon he became Lucifer's first convert and most aggressive spreader of the erroneous teachings through our local system and beyond. He will become the most outspoken, vulgar and sinister follower of Lucifer and he wreaked havoc on everyone he came in contact with.

The Archangel Gabriel

He was the first born of Michael and is the Morning Star of our Local Universe. While he was on inspection of our system, he became aware of Lucifer's intention and went to the Constellation Fathers to determine measures to use in the event of rebellion. They, realizing this was a freewill universe, referred the matter to Michael the Great Prince.

Michael the Great Prince

He is the creator of this Local Universe. He decided, in agreement with the Ancient of Days, not to interfere. He had not finished his required seven bestowals. Therefore, he had not yet been given complete authority. He also hoped the leaders would repent from their erring ways. Because, they were his own children. He also knew that the evil would eventually burn itself out.

During this period each of the colored races fought among themselves and struggled for survival. Lucifer's rebellion and all surviving colors of humanity went through severe regression.

300. Tho. — **200, Tho.**

|---------------------------|---------------------|-----------------------|--------------------------|

Years ago — years ago

6th and final ice sheet

Lucifer launched his Declaration of Liberty:

His orgy of darkness and death and was given ample time to setup and organize before Gabriel would take action.

LUCIFER'S DECLARATION OF LIBERTY
Was stated under THREE HEADINGS:

1. The reality of God the Father.	2. The universe government.	3. The attack on the universal plan of ascendant mortal training.
He declared that God did not exist. He said God was a myth of the creator sons and others to foster fraud on all of creation. He said physical and spiritual gravity were natural phenomena of creation. Personality was not a gift of God the Father. He declared that reverence to God the Father was ignorance. The charges were sweeping, terrible and blasphemous.	**Lucifer declared that local systems should have autonomy. He denied Michael's authority over him. He bitterly attacked the Ancient of Days authority calling them "Foreign potentates" having no authority over him and his domain.** **He believed in the complete home rule if only men and angels had the courage to assert themselves** and **claimed their right.**	He said too much time was spent training ascending personalities. He said the ascenders should enjoy individual Self-determination, paying allegiance to him alone.

At about 200, Thousand Yeas BC. The point the sun began to rise Again (December 25th) became Important to those that worshiped the sun god.They called it the rising **of the sun**.

Periodically a teacher would emerge in various regions. In one race then the other and there would be peace for a time for them. But, it would be lost again and again.

Those tribes close to the prince of the world's headquarters would live in peace until he defaulted and brought darkness and turmoil to all humanity.

200. Tho. **100, Tho.**

|-------------------|----------------------|-------------------------|-------==----------|-Years ago

years ago

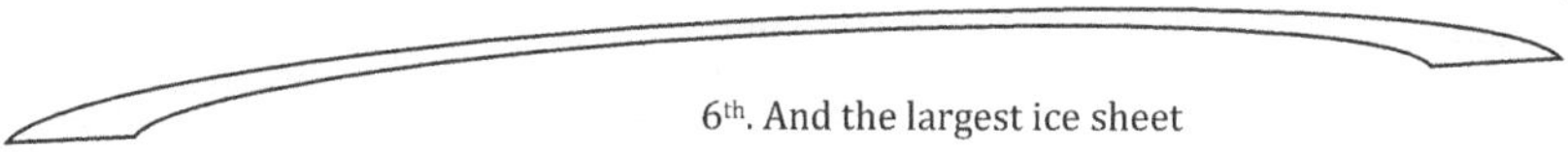

Ice reaches farther point. Great ice tongs or lobes carved out lakes.
Horses, Llamas, tapir and saber-tooth tiger become extinct in North America.

ATLANTIS **civilization spans 200,000 year**

The yellow race developed a devotion to the ONE TRUTH, and they developed peace amongst their tribe.

Lucifer Rebellion: The War In Heaven

LUCIFER'S STAND

Lucifer stood up and declared his declaration of liberty. He denied the existence of God; he said everything was a natural process. Personality was not a gift from God. He claimed the Only Son and the Ancient of Days should have no authority over them in his, Lucifer's local system. He acknowledged Michael as his creator father, but denied his sovereign reign. He asserted that if all would stand up with him they could take authority for themselves. He said he would give the princes of the world complete authority over their worlds and they would owe allegiance to him only.

SATAN'S STAND

Satan stood up at an assembly of the host in the spiritual city of Jerusalem and declared Lucifer as "Friend of man and angels", the god of liberty. Allegiance was to be given to Lucifer only. "SELF ASSERTION" was the battle cry. A new staff was publicly sworn in as Satan brought Lucifer's teachings to the Prince of the World, who was easily persuaded to join the rebellion.

Some of the spiritual sons were beginning to investigate and get involved with the earth and the human race. More and more the various races were regressing because of this involvement.
As the prince of the world was turning away from his responsibilities, most of his staff was moving toward the rebellion.

Some of the staff of the Prince of the World resisted, along with some of the evolving mankind, and fought against this evil slough of the rebellion. They protected the tree of life because it bore the fruit that was needed to sustain the spirit born entities in the physical world. It would have no effect on those born of the physical.

100, Tho. | **75,Tho,** | **50, Tho.**

|------------------------|------------------------|------------------------|----------------------------|

Years ago … **years ago**

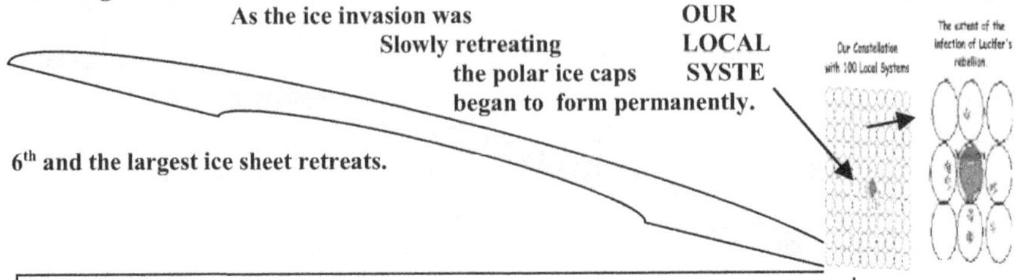

1st. Atlantis disturbance: / Final Lemur submergence.

ESKIMOS

Descendants of the world's Aborigines who were looking for a new home crossed the Bering Strait.

REDMAN MIGRATES

Redman migrates in mass across the Bering Strait into the Americas. They intermingle with early Eskimos.

THE BERING STRAIT SINKS

Isolating the red man and the Eskimos. The red men were aware of the prince of the world's teaching, but lost most of it except a belief in the "Great Spirit".

Atlantis 2nd disturbance the great flood.	10,700 year ago the Final destruction of Atlantis.	**SEA LEVEL FALLING AND RISING** **At 35, Thou. Sea level started dropping. It dropped 137M. (450').** **At 18, Thou. Sea level rose rapidly as the ice age glaciers began to retreat.** **At 11, Thou. Years ago it rose to within 6M (20') close to its present day level.**

Gabriel Fights War In Heaven

Gabriel comes to the Constellation Headquarters and establishes himself on a sphere dedicated to the Father. Those spiritual entities whose loyalty was questioning went back and forth listening to Lucifer and then Gabriel until they made their free will choice.

This war was far more deadly than for the physical destruction. It was a war fought for eternal life.

Some of the ascending souls fought more brilliantly than some of the lower spiritual sons of God who were created for service.

From the time of the declaration of war by Lucifer this entire system was placed on a spiritual quarantine with a complete communication blackout until Lucifer was removed from authority, after which spiritual communication was partially restored.

50, Tho. **40, Tho. (35,797 BC) The default** **30, Tho.**

|-----------------------|-----------------------------|---------------------------|--------------------------|

Years ago years ago

THE REBELLION IN HEAVEN WAS PUT DOWN, BUT THE WAR WAS NOT OVER.

After Gabriel decided that no more would change their mind he, (in agreement with the Ancient of days), and his heavenly host removed Lucifer, Satan and their followers from authority and new or temporary assignments were installed. The rebellion in heaven was put down, but the war was not over. Lucifer and the host that followed him were still able to roam free but without authority. They appeared more often on the worlds that had been led astray by their prince of that world, like ours. The prince of our world had joined the rebellion but, was still in authority.

THE DRAGON

was composed of 2 parties

LUCIFER: Was the head of the dragon breathing fire and destruction.

SATAN: Brought the deception to this and other worlds, he was the deceiver of those worlds.

CALIGASTIA: (the prince of this world) deceived the people he was supposed to lead and guide in this world. He will become known as the Devil. The fallen angels and all the others of the heavenly host that joined in the rebellion are collectively known as the dragon.

WATCHERS Are angels that have been assigned the responsibility to watch over and assist the new life of this world, the sons and daughters of man. Some of these watchers (approximately 200), along with some of the prince of the world's staff, called sons of God, became enamored of the daughters of man and took wives for themselves. Angels are eternal creatures and have no need to procreate. Their offspring were the giants in the world, Spoken of in Genesis.

THE FALLEN ANGELS The fallen angels joined forces with Lucifer, Satan, and the devil and wreaked havoc on the new evolving mankind. They taught the new life how to make corn liquor and fine lining and how to paint their bodies' which in sighted lustfulness. Sex ran ramped and incest led to the point they didn't know their own mother, father or family. They taught them how to make metal weapons of death, make evil witchcraft and sorcery, which led to disrespect for life which led to animal blood sacrifices then to human sacrifices which led to total disrespect for life. They enticed them to worship them. They became false and evil gods. They and their descendants settled in the land of Nod east of Eden.

A FEW SURVIVORS
A few of the sons of God and a handful of early mankind resisted the onslaught and established a new headquarters where they could protect themselves and the tree of life. They began construction of the Garden for Adam & Eve.

ADAM & EVE (35,900 BC)
That was the approximate arrival of the Adamic, material son and daughter, (a spirit born son, and daughter). They are given physical bodies made of the elements and the spiritual breath of life breathed into them. Their bodies glowed with a blue hue. In order to not stand out among the people they wore a hooded full-length robes. That started a custom that still exists today. Their job was to live among the sons of mankind and teach and guide them in their development. Their job was a monumental task. The conditions of this world were nigh impossible to comprehend.

30, Tho. | **25,000** | **20, Tho.**
|-----------------------------|-----------------------|------------------------------|-----------------------|
Years ago | | years ago

ADAM & EVE'S MISSION

They were material sons and daughters of God. (Spirit born and given physical bodies) They were the new spiritual, Social, and biological up lifters. They were also to be the moral rehabilitants of the world.

The World was in shambles. They had some success, but it slowed to a standstill. Eve was becoming impatient. Some of the prince of the world's staff sent one of their fellows to her with the idea that if she would procreate with one of them, Serpintine, they would bring a more advanced son of man into the world which would help speed the process and help the new life advance more rapidly.

EVE WAS BEGUILED

Eve became impatient and began to try, on her own, (without the thknowledge of Adam) to do something at would help. She was misled, tricked, into doing something wrong in order to accomplish something good. The tree of knowledge of good and evil was not a tree at all, but the act of doing evil to accomplish good.

ADAM'S SIN

Adam learned of Eve's actions, he chose to do the same thing motivated by his great love for Eve because he knew she would be denied the fruit of the tree of life and would surely die. The tree of life was brought from the spiritual city of Jerusalem sustained their spiritual bodies indefinitely in the physical world. Adam could not bear the thought of Eve having to go through that experience alone.

ADAM AND EVE'S SIN

Eve's sin was disobedience. Adam was given the plan to follow and Eve was given the conjoint authority. This world was the first world to give conjoint authority to both the male and the female. They were given explicit instructions to never be alone and never act individually. Adam's sin was his choosing to go against God's will, God's plan. Both Adam and Eve's sins were disobedience and choosing to do evil with good intention. Neither Adam nor Eve ever entertained the idea of joining the Lucifer rebellion. They lost their spiritual states as material sons of God and became an elemental man and woman like the sons of man, thus the coats of skin and the necessity to pass through the physical death.

SPIRIT BECOMING MATTER

Because of the spiritual light introduced by the Adamic seed their descendants lived much longer and were much stronger than the normal human beings. That light became dimmer and more distant with each new generation until they too became the norm.

IN THE LAND OF NOD

The Nodite race was descendants of the prince of the world and his staff in the land of Babel later to become Babylonia.

CAINE & ABEL

The enmity between Serpintine and Eve was Cane. Cane having a genetic mixture was different from Abeles. In the first garden, Adam refused burnt offering sacrifices, but they were preferred in the second garden and the land of Nod. Abel would boast and taunt his elder brother, enraging him. Cain's anger rose up and he killed his brother. That same enmity still exists between brothers and religions today. Education, understanding, and love are the only things that will overcome that enmity. Seth, Eves third son carried on with Adam's work.

JARED

Jared led his people to a mountain top where he taught them the ways of the One God. Satan, the fallen angels, and all the evil ones would gather at the bottom of the mountain drinking corn liquor, playing lustful music and having wild crazy orgies. That attracted the attention of Jared's people. He warned them if they went down from the mountain they would not be able to return. When they went down, his heart was broken and he died from grief.

Time Line Of Adam & Eve's Descendants

Genesis:5

0 100 200 300 400 500 600 700 800 900 1,0 1,1 1,2 1,3 1,4 1,5 1,6 1,7 1,8 1,9 2,0 2,1 2,2

| Adam 930yrs. | 930 — **HEBREW MEANING**
& Evw ? — Man
| 130 had
| Seth 912 yrs | 1,042 — Appointed
Yr 130 105 | had
| Enos | 905yrs | 1,140 — Mortal
What yr 235 | 90 had +jh juhyhhhhjgggggg loi89j p. — (will)
Happened | Cainan 910 yrs | 2,105 — All come down to sorrow
To Cain & Able yr 325 | 70 had — (but)
| Mahehaleel 895 yrs | 2,100 — The blessed God
Cain was tiller of yr 395 | 65 had
soil (getting involved) | Jar'ed 962 yrs } 2,233 — Shall come down
in the earth.) yr 460 | at 162 had
Able was a keeper of sheep | Enoch 365 at 965 Walked off with God — Teacher/teaching
(Keeper of higher) yr 622 | at 65 had — (that)
Ideals.) | Methu'selah 969 yrs | 1,656 — His death shall bring down
They represented yr 687 | at 187 | had — (to the)
The duel nature | Lemech 777 yrs | 2461 — Desparing
Of mankind. yr 874 | at 182 | had 1,651
| Noah | 950 yrs | 1,824 — rest (and confort)
Yr 1,056 | had Shem, Ham, and Ja'pheth
before the flood in 1,656.
Date of the flood yr 1,656 | Noah was 200 yrs old |
Time on the Arck Yr 1,656-1657 Approximately (1 yr).
Noah lived 567 After the flood and died yr 1,824 at the age of 950 yrs.

The flood 1.656

Noah with (a famely of 8 people) Plus an untold number of animels.were onboard the Arch for approximately 364 to 370 days depending on a Lunar or a solar year.calculation. That would be very close to One year ± a day or so.

For more information and a deeper understanding of the events that happened before these times Reas the *Urantia book*,*The book of Enoch* and watch *Kile Smith's videos of God in a nutshel.*

Cain's Descendants

Genesis 4: 16-24

Cain went into the LandOf Nod
and knew his wife
Who bore
Enoch
Who begat
Irad
Who begat
Mehujael
Who begat
Lamech (took 2 wives Adah & Ziallah)
Adah bore
Jabal
who begat those who lived in tents
and raised cattle.
Adah also bore
Jubal
who bore musians
Ziallah (Lamech's 2nd. Wife)
bore **Tubalcain** and artist in Brass & iron.
And his sister was **Naamah.**

20, Tho. 15,000 10, Tho.

|------------------------------|------------------------|-------------------------------|------------------------|

Years ago years ago

A MESSAGE FROM CHRIST MICHAEL TO ADAM & EVE

Among other expressions of friendship and comfort, he said "I have given consideration to the circumstances of your default, I have remembered the desire of your hearts ever to be loyal to my Father's will, and you will be called from the embrace of mortal slumber when I come to Urantia unless my brethren decide to call you before then. (They did.) Both were appointed and are now members of the 24 elders.

CHRIST MICHAEL ANNOUNCES HIS INTENTION FOR HIS 7TH BESTOWAL

Unlike his prior bestowals, this time he declared that his 7th and final bestowal would be in this torn and distraught world. The super-material government, the direct spiritual guidance, had been severed for more than 450,000 years.

This announcement was giving those rebels notice that he personally would deal with them and that war would take place in this world we call earth.

God did have a plan by which none should be lost and it was set into motion by this announcement.

OUR WORLD'S MISFORTUNE BECOMES OUR GREATEST BLESSING

The blunders and mistakes of the early world leaders so plunged this planet into a hopeless state of confusion, it was confounded by evil and sin which created a darkness so deep with tangled affairs. It afforded a perfect background for the Creator Son to reveal the matchless love, mercy, and patience of the Paradise Father.

11 GENERATIONS FROM ADAM AND EVE

1: Adam and Eve had
2: Cain and Abel, Cain killed Abel, then they had
3: Seth, a new beginning who had
4: Enos, who had
5: Cainan, who had
6: Mahalalel, who had
7: Jared, who had
8: Enoch, (who walked off with God and was no more) had
9: Methuselah, who had
10: Lamech, had
11: Noah.

Many other children were born in their lineage but are not spoken of.

Much could be said about

THE NODITE RACE

Much could be said about the Nodite race: They were a mixture of some of the surviving tribes of the human species; the prince of the world's staff, and of Cain son of Adam, both were often called the sons of God. Their offsprings were called mighty men of old, the men of renowned, because they were superior physically, intellectually and spiritually. There was also among them the offspring of the fallen angels (watchers) who gave birth to the giants in the world.

BABLOT

A Nodite leader designed a temple and tower as a culture center. The people wanted to build a temple with a tower the likes that had never been seen. One-half wanted it to be a monument to the Nodite people; others wanted it to be a tower of commerce, art, and manufacturing. A smaller group wanted it to be built as atonement for the folly of their progenitors in their joining the rebellion. It was started but stopped. At 12,000 years. a second attempt was made, but it failed and it became known as the land of Babel.

10, Tho.	**5, Tho.**
Years ago	**years ago**

MIGRATION / CONQUEST OF EARLY PEOPLE

Early mankind began to appear on the earth about 1million years ago. The Andonites were more human and less animal mostly in Africa and Europe, then crossed into the Americas as the continents began to drift apart. The Eskimos are the purest of the early races still existing today. Others moved across the European and Mesopotamian arena. They would follow the ice during the ice ages. They moved in all directions with the expanding and contracting ice. At about 500 thousand years ago the six colors of people appeared and began to move north and west intermingling warring or taking mates with the existing mankind. Mankind was bestowed with will dignity. About this time The Prince of the world arrived. He was a material son, a spirit born son with the ability to take up a physical body and exist in both dimensions. He was supposed to be the leader of the peoples of this world and assist them as they developed spiritually, socially and physically. The war in heaven was brought to this world when the prince joined the rebellion and corruption began. Adam and Eve arrived around 35, thousand years ago. They failed by making the wrong choices and their mission was miscarried. Their staff and followers began to intermingle with the existing tribes. The Adamites were a peace loving people and their movements was more of a migration from 35, thousand to about 1 5, thousand years ago spreading across most of the known world.

The prince of the world and his fallen followers became the Nodites and mixed with some of the people and became known as Andites. By this time they became a belligerent race and their movement was more by conquest. From about 8 to 6, thousand they poured out of Mesopotamia into Iran, Egypt, and Turkestan, India, and into China. Their blood line was mixing with almost all the peoples of the known world. The indigo or black race stayed mostly in Africa and was not attracted to the other colors of people. The so-called white race or Arian, developed in the northern, central and southern Europe. The white races, were mostly a mixture of the blue race, and the Andonite blood, with a little mixture of the red and yellow races. They were long-headed, tall and blond.

Somewhere along the line, the Sumerians, Chinese, Indians, Egyptians, and others began to plot the movements of the heavenly bodies. They began to translate them into their worship, and it became a part of their established religions. Astrology was giving birth to Astronomy which will later give birth to Space Science.

Humanity was beginning to think in some abstract terms; art was developing along with mathematics and social ethics. About 7,000 years ago they established the 24 hour day and the 360-day cycle we call the year, and they were able to predict eclipses. As they developed writing, they gave us one of the oldest written stories, "The Epic of Gilgamesh."

MIDWAYERS

Were special created creatures that functioned just above the physical assisting the ascending people on their way to higher spiritual levels. They were a great augment to the Prince of the World and his staff. Unfortunately, most of them joined the rebellion. 42 thousands of them went astray. The remainder was tried, rebels tested and were loyal. They have since grown in number and remain in position. Those that went astray are imprisoned and await trial.

THE CRADLE OF CIVILIZATION

The cradle of civilization was in southwestern Asia, extending from the Nile valley eastward and north Arabia through Mesopotamia, and into Turkestan. Geography and climate were the determining factors. Climate evolution is accomplishing abandonment of hunting and moving into herding and farming. Evolution may be slow but it is terribly effective.

5, Tho. |--| **2, Tho.**
Years ago **years ago**

THE INDUS CIVILIZATION

From 3,000 T0 2,500 was the height of this society. They had public baths with priest rooms for purification. They were a dark skinned female-orientated society with a female deity, the Great Mother, mother earth, mother god with a worship of a fertility or vegetation god. This was analogous to Shiva, a personification of the reproductive power of nature. The father god would come down and mate with the mother god and they would have good crops. The women would pick strong men and mate with them helping the gods. Everyone joined in because they liked helping the gods. That led to the practice of orgies in Babylon, Sodom and Gomorra and Rome. They were a nonviolent group. They were a mixture of the Yellow, blue and orange races.

MACHIVENTA MELCHIZEDEK

About 3,000 years B.C the concept of God had grown very hazy in the minds of man. 1,973 years before Jesus was born, Melchizedek, high Priest of the Most-High God, obtained permission to come to this world. His mission was twofold: 1. Restore and enhance mankind's understanding of God and 2, prepare the way for Christ Michael to come to this world on his 7th and final Bestowal. He first appeared near what was to become the city of Salem, in Palestine. His first appearance to mankind was when he entered the tent of Amdon, a Chaldean herder of Sumerian extraction, and said, "I am Melchizedek, high priest of El Lyon the Most High, the one and only God." After they recovered their composure Melchizedek, with a sweep of his arm said, "El Lyon, the Most-High, is the divine creator of the stars of the firmament and even this very earth on which we live, and he is the supreme God of heaven."

Melchizedek appeared, lived ninety-four years and then disappeared. He never had a need for the tree of life. He became the city of Jebus which later will be called Jerusalem. He established a school. His students learned what they could and took it back to their lands and taught their people, but it became polluted by the existing religions. Lao Tzu and many other great leaders were his students. Abraham went through his school three times and was convinced by Melchizedek to quit his warring ways and follow after the One God.

MELCHIZEDEKS TEACHING THE COMING SON OF GOD

He taught that in some future time another Son of God would come, but that he would be born of a woman; that is why many later teachers said that Jesus was a priest, or minister, "forever after the order of Melchizedek." He taught the people as much as they were capable of receiving. He subordinated everything to the doctrine of the One God, a universal deity, a heavenly creator, a divine Father. They simply could not grasp the idea of getting divine favor for nothing or by faith. They could not grasp the good news that salvation, divine favor, was a free gift to all who would believe and have faith. Abraham halfheartedly believed and that was counted for righteousness. He, Melchizedek, tried to exchange the blood sacrifice of animals for bread and wine, but that was too much for the people of the time. He wore a robe with three concentric blue circles; most people thought they stood for three worlds: men, angels, and God. In truth, they were emblematic of the infinite, eternity, and universality of the Paradise Trinity of divine maintenance and direction.

NOTES:

Worship of the elements gave rise to stacking stones as a witness to an agreement. Disrespect for life gave way to sacrifice of animals then people then virgins to appease the gods the evil ones had established. Mutilation of the body gave way to tattooing and painting the body. Prayer became a process of bartering with God. Making deals with God to make a gain for them. There was very little or no worship involved.

2, Tho. **1.Tho.**

(1) TAOISM 1,700 BC – 1,100 BC Taoism began in the Shang Dynasty which held the belief of the Exalted Being – Ti, and Tien-heaven.	**(2) ARYAN NOBLES INVADE INDUS CIVILIZATION Between 1,500 BC – 1,200 BC The light-skinned Aryan men invaded and married dark skinned Indus female priestesses bringing about the caste system. The word for the caste is "varna" which means color. They spoke Sanskrit and were a male orientated society with very personal gods.**	**(3) RIG-VIDA 800 BC Completed with 1,028 hymns to 33 different gods.** **(5) INDIA'S GOLDEN AGE 500 - 100BC of theology and philosophy.**	**(4) BRAHMAN CAST 700 BC Firmly established as Brahman caste as priest.** **(6) TAOISM 604 – 206 BC Created by Tao-Te-Ching during Han period.**
(7) HEBREW Appeared around 1,400 to 1,200 BC	**(8) JUDAISM 1,005 BC** Kingdom of David established.	**(9) JERUSALEM FELL 597 BC.**	**(10) UPANASHADS MMONOTHEIST 250 BC**
(11) BUDDHISEM 500 BC Birth of Siddhartha.	**(12) YANG & YIN 1,100 BC Developed.**	**(13) GENGHIS KHAN** About 1,200 years ago began his conquest of a great portion of the Asian continent.	**(14) 80 BC** Writing of Pili scriptures for Buddhism.

ENLIGHTENMENTOF MANKIND

For 2,000 years the teachings of Melchizedek was spread around the known world bringing about a better understanding of the One God and preparing the world for the bestowal of the Son of God.

CHRIST MICHAEL'S COMMISSION FOR HIS 7TH BESTOWAL

Prior to his coming to this world, the Great Prince Michael had obtained permission to accomplish his 7th and final bestowal on this war-torn world. That announcement was made throughout the heavens. All those that had followed the rebellion were ready to crush the child's head when he appeared in human form.

He was instructed by the Father what he was to accomplish. His mission was to not only tell the people about God, but to show them God's character and nature through his own actions. He was instructed what to say and do. It was the people's understanding that God was indeed the all-encompassing perfect principle of all creation, but he reigned from afar and distant throne and was unapproachable. Christ Michael's teachings and actions were built around these principles: 1. God in heaven is God the Father of all creation, therefore He is your father and you are His sons and daughters. 2. God is a loving Father, and like any good father, his primary motivation is that unconditioned parental LOVE. 3. Not only does God unconditionally Love you He loves all His children, every man, woman, and child of every race, color and creed on every level, sphere, and dimension of His creation. Since we are all His children we, therefore, are all brothers and sisters. We, therefore, can have a close and personal relationship with our Father in heaven and here on this world. All we needed was to believe and have faith in that truth.

---Taking a break

A Break In The Time Line

To Look At The Conditions In The World

Before starting the story of the 7th BESTOWAL of the Great Prince, Michael as Joshua Ben Joseph, Jesus of Nazareth, we need to look at the world conditions at that time. Civilization was in the best conditions since humanity first appeared on the earth. Rome had unified Europe, built roads, and established a social system. The Greek unified language and culture with a philosophy to a certain degree. The Jewish religion provided a One God religion with moral teachings. Roads connected major cities. The seas were cleared of pirates. Trade and the economy were booming.

The Jews were the most influential group of Semitic people because they possessed a strategic geographic position in the world. After the Jews had been scattered, they built Synagogues throughout the Roman Empire. Those Synagogues were the cultural centers for the new gospel of the Kingdom of Heaven in its beginning. Each synagogue allowed many God-fearing men of gentile belief to attend the temple. These were among the first converts, probably because they were not so enslaved to the Jewish traditions. It is so unfortunate when the Jews didn't realize they were chosen to be the spiritual leaders of the world, not political. It was a sad hour when the Jews refused to learn the good news of the Father's love, which was brought to the world by the Son of God. The Jews did not want to share Yahweh because they were "The Chosen People." They built a wall between themselves and others whom they considered to be heathen, uneducated, unknowing, and unenlightened.

Three languages were predominant in Palestine: 1: Aramaic, which the common people spoke. 2: Hebrew was spoken by the priest and Rabbis and 3: Greek, spoken by the educated and upper class.

From the Greek philosophy, the Gentiles were learning the Epicurean attitude of the pursuit of happiness. Those that developed higher morality did not believe in sensual excesses. That helped the Romans escape from a form of fatalism. It taught that humanity could do something to combat ignorant superstition. The Stoics believed in a controlling reason – fate dominant in all nature. They believed humanity's soul was divine, but trapped in an evil physical body. They believed the soul could attain liberty by living harmonically with nature, but they never found God because they failed to realize, or see, God as a loving Father. The cynic based most of its philosophy from Athens; its religious attitude was based on Melchizedek's teachings. They preached that humanity could save himself if he would. They built fires in the minds creating a spiritual hunger of humanity. In the gentile religions, gods were tribal or national, not personal. Pagan cults were a combination of Hellenic and Latin mythology, patriotism and tradition. The Jews intensely resented emperor worship of the Romans. Astrology was a superstitious pseudo-science of Babylon that developed into a religion in the Greco-Roman empire. The mystery religions promised individual salvation, giving rise to a religious brotherhood which attracted the lower classes. Their services had elaborate ceremonies of initiation and sacraments of worship. Secret Rites and ritual were sometimes gruesome. There seemed to be some similarities between the mystery religions and what Jesus taught, so a word of caution, Mystery religions are based on some mythical legend while Jesus taught on spiritual fact. He brought the bread of life and the waters of life to this world. Paul later will teach moral redemption and ethical salvation. He forsook magic rites and ceremonial enchantment. He taught forgiveness of sin and an endowment of a righteous character of eternal survival quality. Because of the war in heaven which descended to this world, conditions became almost completely lost, Melchizedek kindled spiritual hunger in the hearts and minds of humanity. The world was made ready. Gabriel chose, and the great prince agreed on Mary and Joseph as parents in his, Michael's, final bestowal as a son of man in the elemental universe. He wanted to appear on earth as an average man; this way, the common people might understand him and receive him.

Jewish morals and Greek philosophy were merging into a sustainable religious attitude. At the time of Christ, a short-lived revision of hostilities rose up in Alexandria, but the imperial government fully restored the curtailed liberties of the Jews. No matter where the Jews found themselves, scattered throughout the world by commerce or oppression, and no matter what their theological or philosophic difference was, they all agreed that Jerusalem was the center of their worship and ever looking forward to the coming of the Messiah. The Jews were a stubborn and hard headed people bound by the letter of the law and enslaved by countless traditions. They jealously guarded their God Yahweh and would not share Him with others. Jewish theology was irrevocably settled, forever fixed. Anyone who would dare to challenge those traditions and rituals of legalism, no matter how small, would be rejected. God was unapproachable on high, and to suggest that anyone could have a personal relationship with the Almighty would be accused of blasphemy and put to death.

The Spiritual Development Of Mankind

Early humanity had no concept of God. As they were exposed to the conditions of the world, they began to be afraid of anything they did not understand. That fear drove them to worship the things they feared as gods. In the hopes of appeasing them, they would give something of value to them. Conditions would change by natural events, and that would be interrupted or understood as the gods' acceptance of their offering and changed the conditions. That led to the idea that they had to give something to their God to receive his favor. That then led to the practice of offering sacrifices. As those traditions began to become laws, the people would become entrapped by them. As the Lucifer rebellion began to take effect, those sacrificial requirements became more gruesome, requiring blood sacrifices then live human sacrifices.

The Prince of the world was supposed to help those developing human beings develop a more spiritual nature and teach them about the One God his, humanity's, relations with the One God. He was supposed to teach them morals in keeping with their relationship with one another.That did not happen. Instead, He used their fears to set up an erroneous system of many gods with gruesome requirements that degraded and retarded their spiritual development. In effect, he raped the ones he was supposed to be helping.

Adam and Eve came to this world under extremely difficult conditions with a mission seemingly impossible. When they defaulted, it cast humanity further into the darkness and spiritual confusion, making it more difficult for humanity to find a spiritual relationship with the One God. At this point, it seemed as though humanity was condemned or cursed by God, but humanity has never been condemned, nor cursed by the Father. Humanity has only been loved with unconditional parental love. Humanity was misled and beguiled into making the wrong choices that would have the appearance that God was separated from humanity. The truth is God has never let go of humanity. Humanity, by its choices, has let go of God. Time and time again, God has brought up a great leader to help bring humanity back to the progressive road of spiritual ascension. Never giving too much, but always giving just enough spiritual insight to return humanity to where it should be. Through Melchizedek's teachings and the genetic infusion of the spiritual sons of God, via the staff of the prince of the world and Adam and Eve's family, humanity was being lifted up and put back on track. Those teachings were being spread across the known world and would as always get polluted almost to the point of not being - recognizable. God would again bring up another great teacher to steer them back on course again.

Preparations For The Bestowal Of A Spirit Born Son Of God

Slowly but surely, humanity was being made ready for the next event in the spiritual progression of a world. Every world will have ONLY ONE Spirit born Son that will be born, begotten, into that world as one of that world and will raise the spiritual consciousness of that developing world.

What makes this world different than any other world in our galaxy is the fact that the Creator of our Local Universe, our Galaxy, chose this world for his seventh and final bestowal, where he would deliver the message of the love of the One God for all living creatures within his kingdom. His mission was not only to this world; it was for all worlds within his kingdom. Our world acted like and still, acts like a magnifying glass for all worlds to see and know the magnificent unconditional love of God the Father. He was also finishing his seventh bestowal; each bestowal reflected one of the Seven Master Spirits; thus he was becoming a Sevenfold Trinitized Son of God having complete authority in his kingdom, as the One God has complete authority in all of the creation through the Eternal Triune. Preparations were being made ready in the spiritual dimension for that great event to unfold on this small and insignificant planet we call earth.

The Mindset of Early Civilization

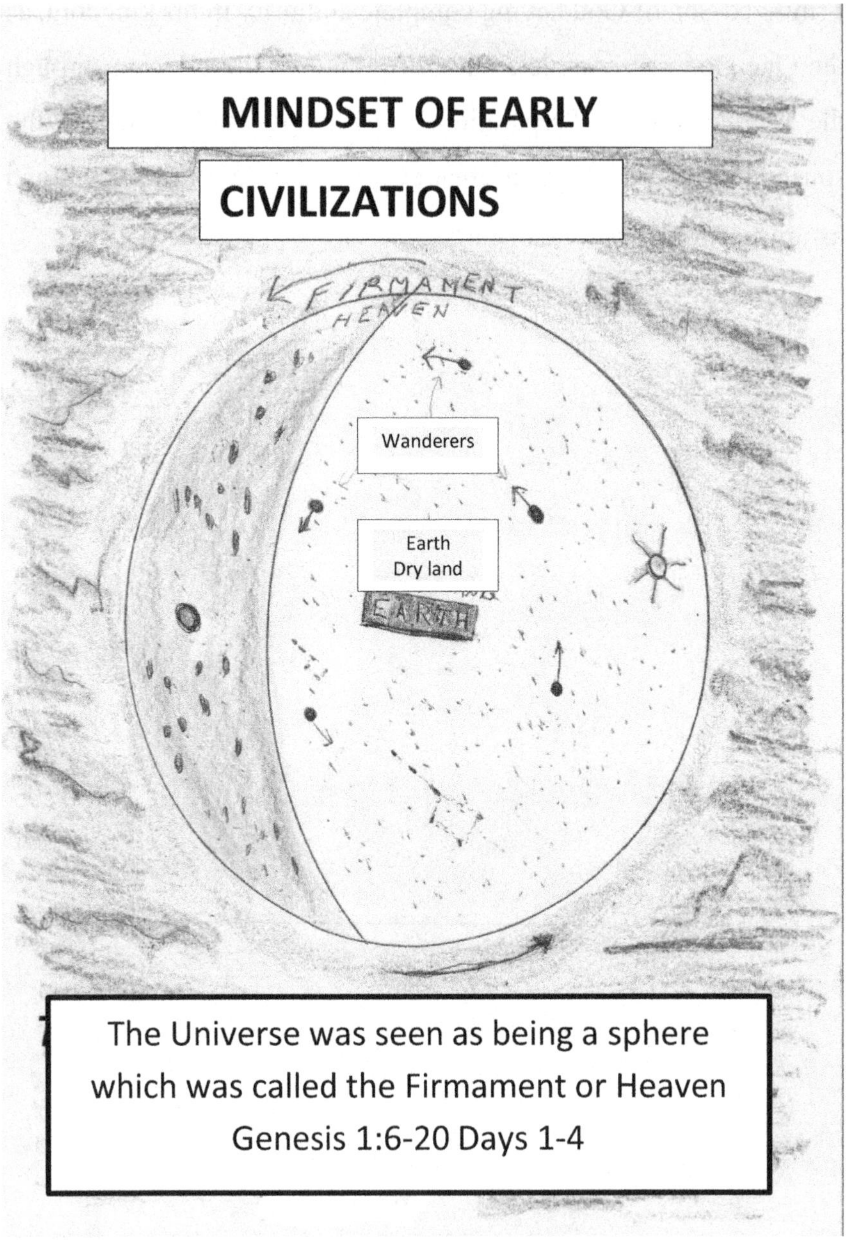

Understanding Early Mankind's Mindset

The universe was thought to be a sphere surrounded by water. It is easier to understand by pushing a balloon or ball underwater in a pool. Water would surround it; in this case, water is an example of the spirit. (God is spirit John 4:24).

The earth was thought to be flat. If you sailed out too far, you would fall off the end of the earth. It wasn't until much later that the earth was shown to be round and was spinning; that caused an uproar in the religious circles. If the earth is spinning, why do I not feel the wind on my face? And what keep people onto the bottom of the earth from falling off? The stars were thought to be attached to the inner surface of the sphere and did not move. We now know that there are other galaxies more or less like our galaxy, but their distance is so far that they Appear not to move.

Then there were the "wanderers," they could see traveling across the heavens, which we now know are planets in motion around our sun. They thought the earth was the center of all creation (geocentric). Then someone had a brainstorm and said what if the sun is the center of creation (solar-centric) and not the earth? Boy, that upset a lot of apple carts. Galileo supported Nicolaus Copernicus theories. That got Galileo ex-communicated from the church. After 350 years, the church reinstated him. I bet he was happy about that since he had long been dead. Somewhere along the line, someone realized the earth was one of the wanderers; we can only imagine the trouble that caused. Their concept was not much greater than the knowledge of our solar system, which is so small and would be insignificant when compared to the 19.6 trillion light-years across our known universe of today. But, humanity is slowly coming out of the dark ages, thanks to our scientific brothers, into a better understanding of Gods magnificent creation which is still unfolding before His very throne. All that sounds like a slow, tedious process of changes. Hum; that sounds like evolution; now that was and still is enough to turn over the apple cart of many religious-minded people. They do overlook the fact that they too are not the same as they were when they were born, nor are they the same as the first day they dedicated their lives to God. All of that sounds like SPIRITUAL EVOLUTION doesn't it. Let us pick up the next piece of the puzzle and study it for a moment.

Mindset of Today

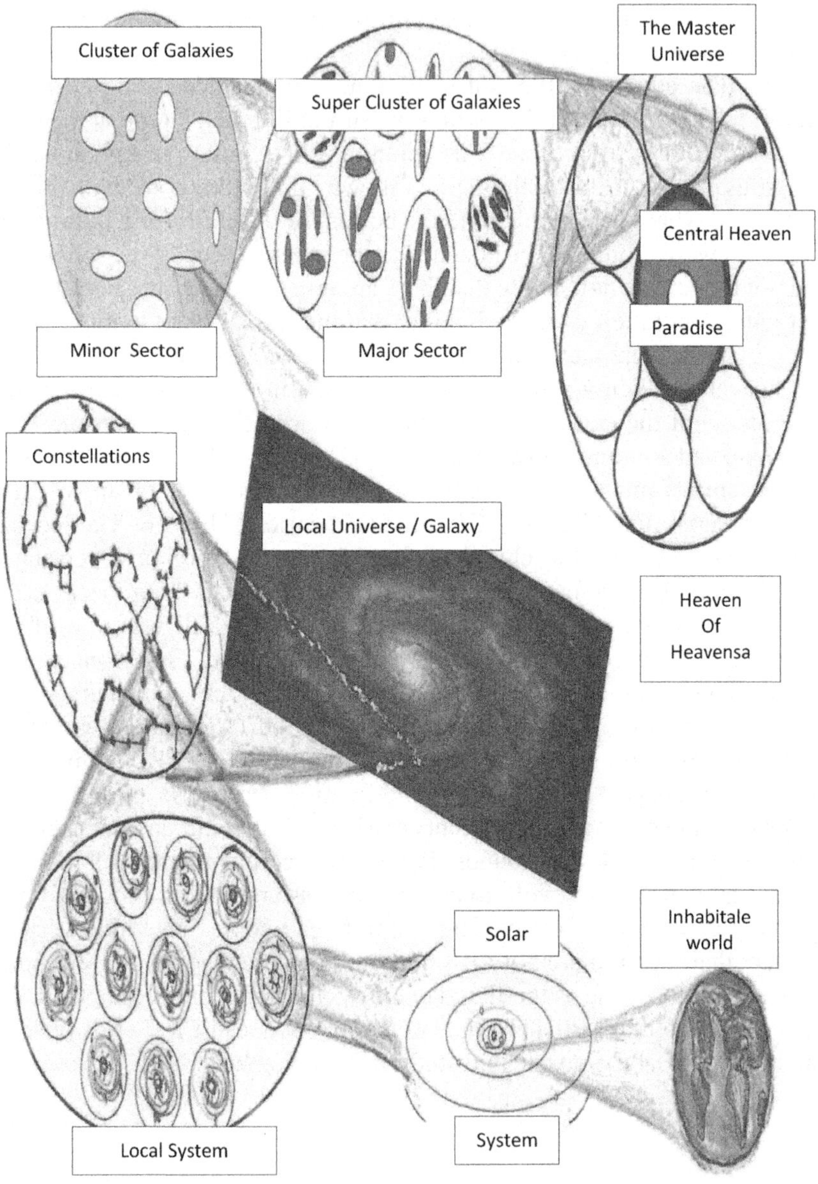

Mindset Of Today

Let us start at the bottom right of the above pictorial representation of our present thinking of creation and where we fit in the picture and our future journey through God's magnificent creation. Our planet (Earth) is the only planet that is inhabitable in our solar system. Our world is fast approaching 7 trillion inhabitants. Other solar systems may have two, three, or four; while a few may have as many as five inhabitable worlds. Our world is spinning at about 1,000 miles per hour. It takes about 365.25 days for our planet to make its journey around the sun. Our solar system is traveling about 220 km/s or 514,000 mph. and takes about 220 million Earth years to make a trip around the hub of our galaxy. That would mean that our solar system has made approximately 20 or 22 trips since it was created 5 billion years ago. Our solar system travels in a group of approximately 500 other solar systems which would be called a local system. There will be assigned a spiritual son of God over the spiritual development for each inhabitable world. His title will be a prince of that world.

A LOCAL SYSTEM: The number of solar systems will vary, what is important is they are supposed to have 1,000 inhabitable worlds when stabilized in the light (spiritually speaking). A spiritual Son of God with the title Son of the Morning is assigned over each of the local systems. He would be the supervisor and administrator over 1,000 princes of those worlds and the entire heavenly host needed to maintain them. Lucifer and his positions are important to us because he is the one that started the war in heaven. His assistant or 1st lieutenant was Satan. He traveled to and fro, and delivered messages to the princes of the inhabitable worlds. It was he who spread Lucifer's erroneous teachings. Local systems are the building blocks of Constellations. There are 100 local systems in each constellation.

CONSTELLATIONS are the building blocks for a Galaxy. There is supposed to be 100 constellations in a Galaxy. Spiritually they are administered by Constellation Fathers, and where they meet is called the mount of the congregation. Often the Sons of the Morning join with them in a variety of meetings.

A GALAXY / LOCAL UNIVERSE is reined over by a Spirit-born Son of God of the order of Michael and is called the Great Prince having creative control and vicegerent authority of God the Father and the Only Son. We of this world know him as the Son of God who incarnated and bestowed himself in the person of Joshua Ben Joseph, Jesus of Nazareth, Jesus the Christ. He is creator, lord, and master of his creation, which is our Galaxy in which we live. We must learn all that his creation has to teach us in our spiritual development. When we do, we will be given his blessing and a new title of a true Son of God with absolute eternal life. We will be allowed to pass through his kingdom into the next level of our spiritual development.

A MINOR SECTOR can closely be related to a cluster of 100 galaxies. It will be like going to college, where we are taught greater spiritual truths. We will grow in our love for God and his great creation. Then we will be advanced to our next level.

A MAJOR SECTOR which is like a supercluster of Galaxies, in that it contains 100 Minor Sectors in it. We continue our higher learning and experiences of God's great Love as we progress through this wonderful school.

A SUPER UNIVERSE contains 10 Major Sectors (or superclusters of galaxies) and is one of seven super universes that makes up the elemental universe which is the Heaven of Heavens spoken of in the Bible. It is also known as the firmament of Genesis. Everything in the heaven of heavens is evolving, going through a slow, tedious process of changing to something higher, better, and greater.

THE MASTER UNIVERSE is made up of the central heaven and the heaven of heavens. The central heaven was created perfect and in a completed state and is not evolving.

PARADISE is the center of all creation and is neither evolving nor moving. It is the permanent abode of God the Father and is wherein we will attain our highest spiritual awareness. Our journeys will culminate when God the Father, the Only Son and the Eternal Spirit allow us to become completely aware of the absolute perfection of the One God.

The Birth, Childhood, And Youth Of Jesus The Young Boy.

The Bible gives us a brief description of these events, but it doesn't give anything about His early life wherein his character and personality developed. Learning about those young tender years as he became conscious of his life's, purpose, and mission is just an awe-inspiring as his mission itself.

During the fifty-five years of research, I have been lead to many different sources of information. I must give recognition to them. Many of the ancient books I have consumed, of course, the Bible and its many translations, the Association for Research and Enlightenment based on Edgar Cacy's readings, Al Miner and Lama Sing (whom I had the pleasure of meeting and receiving a few readings). On occasion, I still hear Lama Sings gentle voice echoing in my mind.

I do believe that the most profound information about Jesus I Have ever read is in the book *Urantia.* There are over 600 pages about this young lad alone. It covers who he was before he came to this world as a physical man and gave his life to humanity. It covers his ministery; it gives an in-depth understanding of his teachings during his life and after his death and resurrection.

In these books, *The story of the Siege*, *The Story of the Legacy,* and *The Story, a timeline* I have followed the pattern of those events, adding a simplified pictorial representation, hoping to make it easier to follow and understand.

The remainder of this book will be devoted to the life of Jesu's birth, childhood, youth, and young adulthood until he heard that John the Baptist was arrested. He promptly removed his carpenters' apron, placed it on the bench, and said, **"My time has come."**

So far, there is a partial timeline of Jesu's early life. This timeline is not complete, but it may give the reader a greater understanding of what Jesu's life was like, what framed his emotions, and thoughts, and understanding. We'll start at 10 BC.

Jan. Feb. Mar. Apr. May Jun. Jul. Aug. Sep. Oct. Nov. Dec.
|10B.C.-------------------|----------------------9-|B.C.------------------------|---------------------8 B.C.

A STUDY OF THE SEGREGATED WORLDS WAS MADE

by the Melchizedeks with Gabriel, and after a study of this report Prince Michael chose this world, Urantia whereon to enact his final bestowal. Gabriel made another visit to Urantia and decided that the Hebrews possessed advantages that warranted their selection for the bestowal. of Michael's approval. Gabriel sent a commission to elect prospective unions. Three were presented, from which Gabriel chose Joseph and Mary, Michael agreed.

GABRIEL APPEARED TO ELIZABETH

About noon, late June, 8 B C Gabriel appeared to Elizabeth. She was frightened by the visitation and did not tell Zachariah for 5 months until she was sure she was pregnant. Beeing married a long time, she had resolved she would not ever have children. March 25th 7 B C John the Baptist was born.

GABRIEL MESSAGE TO ELIZABETH

"While your husband Zachariah, stands before the altar in Jerusalem, and while the assembled people pray for the coming of a deliverer, I, Gabriel, have come to announce that you will shortly bear a son who shall be the forerunner of this divine teacher, and you shall call your son John. He will grow up dedicated to the Lord your God, and when he has come to full years, he will turn many souls to God, and he will also proclaim the coming of the soul-healer of your people and the spirit–liberator of all mankind. Your kinswoman Mary shall be the mother of this child of promise. I will also appear to her."

GABRIEL'S APPEARED TO MARY

8 B C: The day following the conception of Jesus, Gabriel appeared to Mary before Joseph returned home, by the side of a low stone table. After she regained her composure Gabriel presented his message and gave her his blessing.

GABRIEL'S MESSAGE TO MARY

"I come at the bidding of one who is my Master and whom you shall love and nurture. To you Mary, I bring glad tidings when I announce that the conception within you is ordained by heaven and that in due time you will become the mother of a son; you shall call him Joshua, and he shall inaugurate the kingdom of heaven on earth and among men. Speak not of this matter save to Joseph and to Elizabeth, your kinswoman, to whom I have also appeared, and who shall presently also bear a son, whose name shall be John, and who will prepare the way for the message of deliverance which your son shall proclaim to men with great power and deep conviction. And doubt not my word, Mary, for this house has been chosen as the habitat of the child of destiny. My benediction rest upon you, the power of the Most Highs will strengthen you, and the Lord of all the earth shall overshadow you."

-- --

JOSEPH AND MARY WERE MARRIED

March 8 B.C., about three months before Gabriel appeared to Elizabeth. When Joseph was a young man he was hired by Mary's father to build an addition onto their house. At noontime, one day Mary brought Joseph a cup of water and that was when their courtship began. About 2 years later they were married in accordance with Jewish custom. Joseph talked little and thought much and was totally devoted to the teachings of his religion. Mary was cheerful, rarely downcast and possessed an ever positive disposition. Jesus would later display traits from both his parents. Joseph thought he would be a great religious teacher while Mary thought he would become king and sit on the throne of David and make the Jews a great nation.

BIRTH OF JESUS 12 NOON AUGUST 21 7 BC

Jan.	Feb.	Mar.	Apr.	May	Jun.	Jul.	Aug.	Sep.	Oct.	Nov.	Dec.
-7 B.C.											7 B.C.

March 25th John the Baptist was born.

B I R T H O F J E S U S 12 N O O N A U G U S T 21 7 B C

THE BEGINNING OF THE BIG EVENT.

Joseph And Mary had finally settled in their minds that they were chosen to be the parents of the Messiah as they made ready to leave their little home in Nazareth and go to Bethelehem where it has been said the Merssiah would be born

THE TRIP TO BETHLEHEMA

A census was held throughout Rome in 8 B.C., but in Palestine, it was held in 7 B.C. (The Jews refused to be counted with or as the Romans.) Joseph wanted to go alone but Mary wasn't hearing it. She prepared extra rations and everything needed in the event the child was born in route. Early morning August 18 the couple and others departed. They reached Jerusalem before noon August 20 7 B.C. The homes of all their family and friends were already full. Luckily they found a cleaned out grain storage below the Inn, which had been cut out of a rock, in front of a stall and manger with a tent curtain. It was comfortable and thev considered themselves blessed to find it.

The star of Bethlehem: 3 retrograde conjunctions of Jupiter and Saturn in the house of Pisces. The 1st.was May 29th.The 2nd.was on September 29 the 3rd. was December 5th.

Early Aug 18th
Joseph and Mary
Leave Nazareth.
On 20thAug. they arrived around Noon in Bethlehem
JESUS is born Aug, 21st.
August 22nd
Joseph registered birth of Jesus.
29 Aug Jesus was Circum-cised and formally named Joshua (Jesus.
Jesus was 3 weeks old when the Magi visited near Sep. 12th

Zachariah told Simeon
a Judean singer and Anna a poetess from Galilee when Mary would be at the temple. They wrote and sang the hymn of redemption when Joseph and Mary brought Jesus to present their firstborn son to the priest and pay the 5 shekels for his life and pay for Mary's purification from the alleged unclean childbirth.

The Star of Bethlehem
Was a retrograde which involves 3 planets moving around the sun. Fron the Earth it appears that one passes the other, backs up and then moves forward again causing 3 conjunctions of Saturn Represented a king and Saturn, his shield in the house of Pisces, the Jews.

The Hymn of Redemption

Blessed be the Lord, Of Israel, He has visited us and wrought redemption for his people; He has raised up a horn of salvation for all Of us. Even as he spoke by the mouth of his holy Prophets. Salvation from our enemies and from the hand of all who hate us. To show mercy to our fathers, and remember His holy covenant, the oath which He swore to Abraham our Father, To grant us that we, being delivered out of the hands of our enemies should serve him without fear, in holiness and righteousness before Him. All our days.

Yes, and you child of promise, shall be he called the prophet of the Most High; For you shall go before the face of the Lord to establish His kingdom; To give knowledge of salvation to His people. Rejoice in the tender mercy of our God because the dayspring from on high has now visited us. To shine upon those who sit In darkness and in the shadow of death, to guide our feet into ways of peace. And now let your servant depart in peace, O Lord, according to our word, for my eyes have seen your salvation, which You have prepared before the face of all people; light for even the unveiling of the gentile and the glory of Your people Israel.

Joseph and Mary were completely surprised

Joseph and Mary stayed in the inn 3 weeks, moved in with family in Bethlehem 1yr

That conjunction told the Mesopotamian priestess who was in Babylonia, who were astrologers and knew the Jews were looking for a messiah who would be king of the Jews. They were looking for this legendary "Light of Life." The Magi, those wise men from the east, first went to Herod looking for this King of the Jews. Then Zacharias met them and directed them to Jesus who was now almost 3 weeks old.

WHY NOT 25 DECEMBER

1: Errors in calculating the time for the calendar over long periods of time setting B.C. / A.D. 7 years off. 2: No good sheepherder would expose his herd to the cold of December in Bethlehem. 3: The census timing. The Jews would not allow themselves to be counted in a Roman census; it was held one year later, And 4: the retrograde of the wandering stars (planets). NOTE: The date and time of His birth are not as important, as the fact that he was born. He was the only spirit born son of God to also be born in this world, a son of God and a son of man; thus the only begotten son of God.

WHY 25 DECEMBER

The sun god, Sol Invictus, was a degeneration of the Ahura-Mazda deity concept of Zoroastrianism. Mithras was a warring champion in his struggle with the god of darkness not only for himself but for all others. Their greatest feast was December 25th. That was the rising of the sun. They would eat the flesh and drink the blood of the sacrificed animals to ensure their survival after death. They were very strongly devoted and it was difficult for early Christians to convert them, until they accepted December 25th as the rising or birth of the Son as Jesus birthday. That won the hearts and souls of all of the **sun** worshipers over to the **Son** worshipers.

Jan. Feb. Mar. Apr. May Jun. Jul. Aug. Sep Oct Nov Dec.

|6-B.C.----------------|------------------ Aug. 21 Jesus 1 years old. ----------------6|B.C.

HEROD ACTS **Herod was not satisfied with the answers of the Chaldeans priest; therefore he sent his watchers to gather information. They brought him part of the hymn of redemption, but they failed to follow Joseph and Mary. That enraged Herod so he made them search for more than a year and they found nothing. That enraged Herod even more. Then he decided to end the matter by killing all boy babies under the age of 2.**

JOSEPH & MARYIN HIDING Joseph and Mary lived in hiding with their families in Bethlehem. Joseph was afraid to look for work in fear they would follow him to his family.

JOSEPH & MARY DEPARTFROM BETHLEHEM There was some believer in the coming of the Messiah in the service of Herod and they told Zacharias of Herod's plan to kill the child. Joseph was also told and led by a dream and the night before the massacre they packed up and left for Alexandria in Egypt.Joseph's funds had run out so the trip was funded by Zachariah Jesus was near one and a half years old when they left for Egypt.

The Massacuue of 16 boy babies

Herod hoped to make sure that the child who was to become "king of the Jews" would be destroyed. In the middle of October, 6 B.C. he prepared a directive that a systematic search be made of every house in Bethlehem and that all baby boys under 2 years of age should be killed. Intrigue and murder were no stranger in the house of Herod. It was cruel, but it was not looked upon as strange. Even some of his own family was killed because of his paranoia. Thus 16 boy babies perished in Bethlehem of Judea. the house of Herod. It was cruel, but it was not looked upon as strange. Even some of his own family was killed because of his paranoia. Thus 16 boy babies perished in Bethlehem of Judea in the house of Herod.

Jan. Feb. Mar Apr May Jue Jul Aug Sep Oct Nov Dec.
|5-**B.C.**----------------|----------------------|----------------------|-------------------**5**|**B.C.**

Sojourned in Alexandria, Egypt for two full

While in Alexandria, Joseph, Mary and little Joshua stayed with well-to-do relatives of Joseph's family.

Mary, Because of the uncertainty in Bethlehem and the traveling, did not wean the baby until they arrived safely in Egypt and were settled down to a normal life.

Mary maintained one long and constant vigil lest anything befalls her or her son in any way that would interfere with his future mission on earth. Only an affectionate mother can know the burden that Mary carried in her heart for the safety of her son during these years of his infancy and early childhood.

Joseph found work immediately and was able to provide for his family. He was hired as a carpenter, and then promoted to Foreman over several workers on a public building. He later decided he would become a builder when he returned to Nazareth.

Joshua had two little friends his own age and about six others that were close to his age as friends.

Mary wanted to protect Jesus, and not let him play so much with others, but Joseph and friends counseled Mary that it might deprive Jesus of the experience of adjusting to children his own age and become self-conscious. Mary agreed but made it her business to always be on watch while the little ones played.

Aug. 21 Jesus 2 years old.

VISIT BY ALEXANDRIA BELIEVERS FROM MEMPHIS

Aside from a few friends and relatives, no one was told about Jesus' being a "child of promise." One of Joseph's relatives told a few friends from Memphis, who were distant descendants of Ikhnaton. They and a few Alexandrian friends came to visit the child and pay their respect and hopefully convince them to stay in Egypt. They suggested that Jesus would be able to have a greater influence on the world from Alexandria than from some distant place in Palestine.

Their efforts delayed Joseph and Mary's departure for Palestine for some time after they received the news of Herod's death in 4 B C.

It was not until all their efforts failed, and their invitation declined, that Joseph was presented a complete copy of the Greek translation of the Hebrew Scriptures for Jesus. The textbook was a complete version of the law and the prophets, including the Psalms.

Jan. Feb. Mar Apr May Jue Jul Aug Sep Oct Nov Dec.
4-B.C.---------------|---------------------|---------------------|-----------------4-B.C.

Depart Alexandria, Egypt. Arrive At Port Joppa Aug. 4 B.C.

Jesus enjoyed good health and continued to grow normally, like any other child during the two years in Egypt.

Leaving Egypt Joseph and Mary finally took leave of Alexandria on a boat belonging to their friend Ezraeon, bound for Joppa. They arrived at that port late on August 4 B.C.

Aug. 21 Jesus 3 years old.

Jesus was about 3 years and two months old when they arrived at Nazareth.

Back in Bethlehem
They went directly to Bethlehem, where they spent the entire month of September in counsel with their friends and relatives trying to decide to stay there or return to Nazareth.

The debate in Bethlehem
Mary had never given up the idea that Jesus ought to grow up in Bethlehem the city of David. Joseph did not really believe that their son was to become a kingly deliverer. Mary thought it to be the ideal place for the new candidate for David's throne to be reared. Joseph had great fear for the child in Bethlehem or any other city in Judea. He would rather take chances with Herod Antipas rather than with his brother Archelaus, Joseph reasoned would be more likely to follow his father's menacing policies.

Back in Nazareth October 4 B.C.

The return home
Joseph was outspoken in his preference for Galilee as a better place in which to rear and educate the child, but it required three weeks to overcome Mary's objections.
In early October 4 B.C., they departed from Bethlehem for Nazareth going by way of Lydda and Scythopolis. Joseph's family members would not hear of them making the trip alone. It was unsafe for two people and a small child traveling to Galilee by Jerusalem and the Jordan valley, nor was the western route safe either. Five family members and Joseph set out afoot with Mary and the child riding on the back of the new beast of Burden. On the fourth day of the journey, they reached their destination in safety.

Jan Feb Mar Apr May Jun Jul Aug Sep Oct Nov Dec.
|3-B.C.--------------|------------------------|----------------------|------------------ 3 B.C.

Home sweet home

Their departure from Alexandria and arrival in Nazareth was completely unannounced. Joseph's married brother had occupied the home and was surprised to see them. He moved his family out the next day and Mary for the first time since Jesus' birth, (2yrs, and 8 and ½ mos.) settled down with her little family to enjoy life in her own home.

Jesus was in good health and had a very active mind like most 3 and 4 year-old boys.

APRIL 2nd 3 B.C.
Birth of James, Joseph and Mary's second son.
Jesus' brother.

Midsummer Joseph builds a small workshop near the village spring and the caravan tarrying lot. As Jesus grew up he would spend about half his time helping his mother and the other half he would spend in the workshop working and listening to the gossip from travelers from all corners of the world.

Baby brother James
Jesus was thrilled about having a baby brother. He would stand around for hours observe the babies early activities.

JESUS PLAY TIME
consisted of playing in the back of the shop with the shavings and blocks of wood found around the shop with his new and loyal friend, the son of a nearby neighbor.

Aug. 21 Jesus 4 years old.

JOSEPH'S BUSINESS
Was picking up and expanding. He had two of his brothers as assistants and a few workers that were sent out on a variety of construction jobs around the area. In the shop, he began to work making yokes and plows and other wood works. He also did work with ropes, leather, and canvas.
Occasionally Joseph would take Jesus on some of the nearby work with him.

OUTBREAK OF INTESTINAL TROUBLE

In July there was an outbreak of malignant intestinal trouble that spread all over Nazareth. Mary bundled both her children and fled to the country home of her brother which was several miles south of Nazareth on the Megiddo road near Sarid. They stayed there for two months. Jesus greatly enjoyed this, his first experience on a farm.

AGREEMENT BY JOSEPH AND MARY

Joseph and Mary both agreed not to tell the people around Nazareth that Jesus was a child of promise.
They agreed to refrain from all mention of these matters to anyone. And they were both very faithful in keeping this promise.

Jan Feb Mar Apr May Jun Jul Aug Sep Oct Nov Dec
2-B.C.--2 B.C

COMING OF JESUS' THOUGHT ADJUSTER

A little more than a year after returning to Nazareth (February 11, 2 B C) the boy Jesus arrived at his first personal and wholehearted moral decision; and there came to abide with him a Thought Adjuster, a divine gift of the Paradise Father.

Jesus was no more aware of the coming of the divine Monitor than are the million upon millions of children before or after that time who likewise have received these Thought Adjusters to indwell their minds and work for the ultimate spiritualization of these minds and the eternal survival of their evolving immortal souls.

JESUS' THOUGHT ADJUSTER

The Thought Adjuster that worked with Jesus was the same that served with Machiventa Melchizedek, thus gaining the experience of functioning in connection with the incarnation of a super mortal being living in the likeness of mortal flesh.

PRIOR TO THE THOUGHT ADJUSTERS

The responsibility of Christ Michael was under the direct supervision of the Universe Rulers up to this point. After the arrival of the Thought Adjuster, their responsibilities were terminated and that responsibility was turned over to his personal Thought

AFTER THE THOUGHT ADJUSTER

From this point throughout the human unfolding of the incarnation, the guardianship of Jesus was destined to rest in the keeping of this indwelling Adjuster and his associates as directed.

Birth of Mariam Jesus' sister July 11, 2 B C

Jesus greatly enjoyed his baby brother and little sister and he was a great help for Mary in their care.

Aug. 21 Jesus 5 years old.

THE TRADITION OF REARING A CHILD IN NAZARETH

A child's life divided into seven stages.

1. **The newborn child, the first eight days.**
2. **The suckling child.**
3. **The weaned child.**
4. **The period of dependence on the mother, up till the end of the fifth year.**
5. **The beginning of independence of the child and, with sons, the father assuming responsibility for their education.**
6. **The adolescent youth and maidens.**
7. **The young men and young women.**

JESUS'S 5TH BIRTHDAY

Jesus had reached the 5th stage of child rearing according to the Galilean Jews child's career On August 21, 2 B C Mary officially turned Jesus over to Joseph for further instructions. Of course, she continued her interest in the child's learning, teaching him the care of the vines and flowers around the house. She also provided him with a summer bedroom on top of the house, and a sandbox. There he could practice drawing maps and writing Aramaic, Greek and later Hebrew; he became very fluent in all three languages.

Jan Feb Mar Apr May Jun Jul Aug Sep Oct Nov Dec.
1-B.C.-------------|-----------------------|---------------- ------|-------------------1 B.C

Jesus, inquisitive mind
Other than a few minor childhood diseases Jesus was a perfect child with a highly inquisitive mind. He sought to understand everything around him. there came a line of questions, his father and mother were kept busy trying to explain and talk them out with him. Jesus was astonished when Joseph told him, "My son, I really do not know." When Jesus asking about the cause of a minor earthquake recently felt.
Joseph was cautious not to say it was caused by God or the devil, he knew that Jesus would have a stream of questions to follow which he could not answer. He had already learned that Jesus would continue asking more and more questions until he had a satisfactory answer.

Zacharias, and John Visit. Early summer brought these two families together for a few days. It was a great time for Jesus and John. They played in the sand box and with the blocks on the roof of the house. The adults had a pleasant time talking over their ideas of the future of their two sons. John was from near Jerusalem where his father was a priest in the temple. After John left Jesus began to develop a great interest in the history of Israel and began to inquire a great deal about the Sabbath rites, the synagogue sermons and recurring feast of commemoration.

Aug. 21 Jesus 6 years old.

Joseph explains the Sabbath
1st The midwinter festive illumination: It lasted eight days, starting with one candle the first night and adding one each night. This was dedication of the temple after the restoration of mosaic services by Judas Maccabee. 2nd The early springtime celebration of Purim: The feast of Ester and Israel's deliverance through her. 3rd The solemn Passover: The children at home always remembered it as the time they ate no leavened bread the whole week. 4th The feast of the first harvest. 5th. The most solemn the feast of the Jewish New Year, the Day of Atonement.

At first, Jesus' young mind did not fully understand these rites, but as he did he fully entered into all the feast of the tabernacles.

Jesus deciding on his career
Joseph and Mary told only a few close friends, about Jesus being a child of destiny: Jesus over heard Joseph and Mary discussing Gabriel's message. Mary thought he would be a king like King David. Joseph thought he would be a great teacher or leader. Jesus often thought he would become a carpenter, a fisherman, or a potter. He liked arts and craft. Jesus and his friend knew a potter who liked competition, that was fun for the boys. John's vised then he consider becoming involved with the synagogues. Joseph, and Mary would be befuddled with Jesus, he would say he wanted to have a little talk with his father, instead of sticking to the traditional prayers. He obeying them praying those prayers then he have his little talk with God. Mary, a loving mother and a strict disciplinarian. Joseph, was more patient and talked things out with him. Once Jesus understood, he would cheerfully join in and follow his advice.

Jesus recieved custody of the text
A Greek translation of the Hebrew Scriptures, including the law, the prophets, and the Psalms. It was the only complete copy in all of Nazareth. A steady flow of sincere searchers for the truth visiting Joseph and Mary's home. Jesus enjoyed listening to them. On Jesus 6th birth day he was given custody of this sought after textbook. He was told that it was a gift for him from friends from Alexandria. Jesus was delighted.
Mary taught him the Nazarene dialect. Joseph taught him Aramaic, which was spoken by the traveling caravans. Now he was learning Greek. Joseph used this text as a study guide. Jesus had a sand box. He used it for writing and drawings and maps. Jesus didn't understand why some of these were disapproved by the Rabbis.

Jan. Feb. Mar. Apr. May Jun. Jul. Aug Sep. Oct. Nov. Dec.
1-A.D.---1 A.D

THE CITY OF NAZARETH

Nazareth was a caravan way station and was a crossroads of travelers. It was largely gentile in population. The Jews in Nazareth were widely known as the center of liberal interoperation of Jewish traditional law. In Galilee, they mingled freely with gentiles more than in all of Nazareth. Their interpretation of the social restrictions based on the fears of contamination as a result of contact with the gentiles; thus, prompting the common saying, "Can any good thing come out of Nazareth."

Jesus was such a good student that he was given one week off each Month.

He would spend one week visiting his fisherman uncle on the shores of the Sea of Galilee, Near Magdala.

One week he would work on the farm of another uncle (his mother's brother). It was about 5 miles south of Nazareth. He was learning a lot about many of the various trades of people of his time.

There was a friendly competition between his aunts and uncles for his company.

March 16th A .D. 1 The 4th member, of the family, **Joseph**, was born.

Playtime for Jewish children

During those times playtime for children was mostly at weddings and funerals, which they saw far too often **and** were so spectacular. They sang and danced but had few organized games, unlike children of later days who had so much enjoyed.

In August Jesus enters school

Jewish children were supposed to begin their formal education in the synagogue. The lad was already fluent in Aramaic and Greek now he would acquaint himself with the Hebrew language. For 3 years he will attend the elementary school of the Nazareth synagogue. Here he studied the rudiments of the book of the law as it was recorded in the Hebrew tongue.

Aug. 21 Jesus 7 years old.

Jesus birthday text

As Jesus was entering school the Jews had just inaugurated a compulsory educational law. It was customary for the pupils to choose their "birthday text," sort of like a golden rule to guide them through their studies. Jesus chose from the prophet Isaiah: "The spirit of the Lord God is upon me, for the Lord has anointed me; he has sent me to bring good news to the meek, to bind up brokenhearted, to proclaim liberty to the captive, and to set the spiritual person free."

Jan. Feb. Mar. Apr. May Jun. Jul. Aug. Sep Oct. Nov. Dec.
2-A.D--2 A.D

This year Jesus learned how to milk a cow and care for the other animals. During this year, and the next year he learned to make cheese and weave. By the age of 10, he was an expert loom operator. He and his friend Jacob were getting interested in pottery as a possible profession. It was fascinating to watch Nathan mold the clay with his hands as he challenged the lads to do the same.

THE METHOD OF TEACHING AND LEARNING

At Nazareth, the pupils sat in a semi-circle on the floor while the teacher, the chazzan, an officer of the synagogue, sat facing them. Their studies began with the book of Leviticus, then the books of the Law, followed by the study of the Prophets and the Psalms. The synagogue had no text books. The chazzan would speak the words and the children would in unison repeat it after him. This would be repeated over and over until the children had memorized the scriptures. When they had access to the written word the students would learn by reading aloud to the class.

School was getting difficult, for the teachers

The school was great, but not smooth. Jesus asks too many questions especially about religion and science, particularly about geography and astronomy. Why are there dry seasons and rainy seasons in Palestine? Why was there a difference in temperature between Nazareth and the Jordan valley? He never ceased to ask such intelligent but perplexed auestions.

Questions, questions, and more questions; there was no end in sight for the teachers.

Friday evening April 14th Jesus' 3rd brother, Simon was born,

JESUS WAS A GREAT HELP TO HIS MOTHER

Jesus was now old enough to be of great help to his mother and the children. He enjoyed being their big brother and their teacher. They always enjoyed themselves when he spent time with them telling them stories, he was a great story teller, and they love it. Their love for each other always brought happiness to Mary as she maintained a watchful eye over her growing family.

JOSEPH EXPANDS BUSINESS

In June Joseph turned the shop over to his two brothers and entered into his work as a builder. Before the end of the year, his income tripled. Never again, until after Joseph's death did this Nazareth family feel the pinch of poverty. As their family grew their income grew. They were able to spend more money on education and travel.

Aug. 21 Jesus 8 years old.

JESUS BEGINS TO MEET AND LOVE STRANGERS

On occasion, Jesus went on trips with his earthly father to nearby cities on construction jobs. Jesus enjoyed meeting different people from all around. He developed an understanding of their ways of life and a sincere love for people.
Jesus having such an inquisitive mind ask many questions, and then listen intensely as they explained things to him. He never passed up an opportunity to tell those stories which he so greatly enjoyed. Jesus' love for mankind was developing and at the same time as those people got to know this lad, it was hard not to love him in return. A stranger is a friend you haven't met yet.

Jan. Feb. Mar. Apr. May Jun. Jul Aug Sep Oct. Nov. Dec.
3-A.D---3 A.D

SNOWFALL

Early January a great snowstorm occurred in Galilee. It was two feet deep, the heaviest snow fall in all of Jesus life time.

JESUS FREE TIME AND EXTRA LEARNING EXPERIENCES

About this time Jesus met a teacher of mathematics from Damascus, from whom he learned a lot of new techniques about numbers and proportions and distances.

JESUS WAS VISITED AND OFFERED AN OPPORTUNITY TO ATTEND A JERUSALEM ACADEMY

In February, Nahor, a teacher in a Jerusalem academy of the rabbis, came to observe Jesus after visiting Zachariah who insisted he visit Nazareth to see Jesus. At first, he was setback because of Jesus frankness and unconventional manner of relating himself to things religious. But he assumed it was because of the remoteness of Galilee from the centers of Hebrew learning. He suggested that Jesus be allowed to go with him to attend the academy. Mary thought it would be good for Jesus and be of help in becoming the messiah, the Jewish deliverer that she had hoped he would become. Joseph, on the other hand, thought he would not become the king, but he was sure he would fulfill a destiny. What that destiny was he was equally not sure of. Because of their differences, Nahor requested permission to lay the whole matter before Jesus. Jesus listened attentively. After talking with Joseph and Mary and his friend Jacob the stone mason, the father of his favorite friend. He was still not sure what he should do, so he decided to have a little talk with his Father in heaven. He was not perfectly sure of the answer; but, he felt he should stay at home. He said, "with my father and mother," adding, "they who love me so much should be able to do more for me and guide me more safely than strangers who can only view my body and observe my mind but can hardly truly know me." Every one marveled at his answer. Nahor went back to Jerusalem and the subject of Jesus leaving was not brought up for many years.

THE RESULTS OF JESUS' CHOICE

In Alexandria, Jesus would have been taught by Jews and along exclusively Jewish lines. At Nazareth, he secured an education and received training which more acceptably prepared him to understand the gentiles, and which gave him a better and more balanced idea of the relative merits of the Eastern or Babylonian, and the Western, or Hellenic, view of Hebrew theology.

Jesus' only accident

The only real accident in Jesus life occurred this year. In July he was playing on the rooftop when a sudden sandstorm blew in from the east. The hot winds, carrying blast of fine sand, which usually blew during the rainy season in March and April. As Jesus tried to get off the roof via the steps on the side of the house he fell. He was not injured, but Joseph wasted no time installing two hand rails up the steps.

Aug. 21 Jesus 9 years old.

JESUS 2ND SISTER, MARTHA
WAS BORN Thursday night September 13th.Joseph & Mary's 6th child.

JESUS DEVELOPS A LOVE FOR MUSIC

Jesus made arrangements to exchange dairy products for lessons on the harp. He had an unusual liking for everything musical. He later did much to promote an interest in vocal music among his youthful associates.

Jan. Feb. Mar. Apr. May Jun. Jul. Aug. Sep. Oct. Nov. Dec.
4 AD--4 A.D

JESUS GETS INTO TROUBLE

Jesus loved to draw landscapes, pictures, images and maps. He also loved creating all sorts of clay models. This winter the most serious trouble for Jesus came up because he dared to challenge the chazzan regarding all of these as being idolatrous in nature. Everything of that sort was strictly forbidden by Jewish law. The trouble really began when a student found a charcoal drawing, a picture of the teacher; there it was, as plain as day and many of the students saw it. The committee took the lad and demanded that Joseph does something about his eldest son's lawlessness. This was not the first time they had complained about the doings of their versatile and aggressive child.

JESUS DEFENDS HIMSELF

Jesus sat on a stone outside the back door listening to the indictment of his artistic efforts for some time. He resented their blaming his father for his alleged misdeeds, so in he marched! Fearlessly he confronted his accusers. The elders were thrown into confusion; some thought it was funny and other were indignant and thought he was sacrilegious if not blasphemous. He finished with self-control stating that he would abide by all his father's wishes in all matters controversial as long as he lived with him.

THE COMMITTEE DEPARTED

They left in silence. Joseph felt compelled to follow the rabbinical interoperation of the second commandment should prevail.

Even though Jesus was unconvinced of wrong doing, he never again drew or molded anything in the likeness of anything as long as he lived in his father's house.

Giving up this favorite pastime was one of the great trials of his young life.

Jesus was an original thinker and was in constant collision with the so called "oral law". He always adopted himself to the practice of his family. He was a leader even as a lad. Organizing groups and playing activities with others.

Jesus journeyed with his father to the top of the summit of Mount Tabor. From there on this clear day, this young lad felt as though he could see the entire world, except India, Africa, and Rome.

Jesus becoming aware of his mission

July 5 Jesus began to express feelings and ideas which indicated he was becoming self-conscious of the unusual nature of his life mission.

Aug. 21 Jesus 10 years old.

Jesus enters the advanced school of the synagogue.

Jesus moves into higher learning

Jesus graduates from the elementary school of Nazareth and began his studies in the advanced school and committed to memory, by the method of repeating aloud, the deeper teaching of the sacred law for 3 years. Jesus questioned everything, for some it was upsetting, but for others, it was a refreshing change and a challenge to answer his questions.

Jesus helps his uncle with the harvest. He was fast learning something about everything.

Joseph got busy building an addition which would be a new bedroom and workshop in which he built a workbench for Jesus and gave him his own tools. Jesus became very efficient at making yokes.

Jan. Feb. Mar. Apr. May Jun. Jul. Aug. Sep. Oct. Nov. Dec.
5-A.D-------------|--------------------|-------------------|---------------5 A.D

This would be Jesus last year that he will enjoy free play and youthful joyousness. Events will bring him into manhood before what would be considered to be the norm for his time.

JESUS RECEIVES ADDITIONAL INSTRUCTIONS ON HEBREW SCRIPTURES.

The Chazzan spent one evening each week with Jesus, helping him to master the Hebrew Scriptures. He was greatly interested in the progress of his promising pupil, but it was difficult for him to understand Jesus lack of interest in going to Jerusalem to continue his education.

JESUS FEELS THE STING OF JOSEPH'S PERSONAL INDIGNATION

In the middle of May Jesus accompanied his father on a business trip to Scythopolis, the chief Greek city of the Decapolis, the ancient city of Beth-Shean. Joseph recounted much of the olden history of King Saul, the Philistines, and the subsequent events of Israel's turbulent history. Jesus was tremendously impressed with the clean appearance and well-ordered arrangement of this so-called heathen city. He admired the beautiful marble temples dedicated to the worship of the "heathen" gods. Jesus had often gazed upon this distant city from atop the mountain. His father tried to distract the lad by commenting on the beauty of the Jewish temple. Jesus liked the idea of sports and playful competition. That night he suggested to his father that they should go back home and setup a program in Nazareth for all the Jews.

This was the only time Jesus ever saw his father angry with him. When Joseph heard his first-born son express such un-Jewish sentiments he forgot his usual calm demeanor and seized Jesus by the shoulder, and angrily exclaimed, "My son, never again let me hear you give utterance to such an evil thought as long as you live!"

Jesus was startled by his father's display of emotions; he had never before been made to feel the personal sting of his father's indignation and was astonished and shocked beyond expression. Jesus' only reply was, "Very well, my father, it shall be so." Jesus never again alluded to the games as long as his father lived.

Later on, Jesus saw the Greek amphitheater at Jerusalem and learned how hateful such things were from the Jewish point of view. He did, however, encourage playful competition with his disciples and friends.

AD 5
Wednesday
June 24th
Jude was born, Joseph and Mary's 7th child. Because of, complications Mary was very ill and several weeks Joseph stayed home.

Aug. 21 Jesus 11 year Old.

JESUS LEARNS ABOUT WORLD AFFAIRS.

Jesus spent considerable time at the caravan supply shop, and by conversing with the travelers from all parts of the world, he acquired a store of information about international affairs that is amazing considering he was only 11 years old. He greatly enjoyed their conversations and they likewise enjoyed his keen mind and his pleasant manners.

Jesus spent much of his time running errands for his father and began to assume more responsibilities around the home and at work.

Jan. Feb. Mar. Apr. May Jun. Jul. Aug. Sep. Oct. Nov. Dec.
6-A.D------------|--------------------|-----------------------------------6 A.D

JOSEPH AND MARY HAD DIFERENCE OF OPINIONS AS TO JESUS' DESTINY

Jesus becomes more acutely aware of the different opinions that Joseph and Mary had about his being a child of destiny.

Mary believed that Jesus would become a king like King David, and make the Jews into a great nation among nations, while Joseph believed he would be a great leader or prophet and teacher.

Neither of them was ever made aware that he was the Creator Son of our local universe incarnate in the flesh of this world.

JESUS DECIDING NOT TO BECOME A KING

On occasions when Jesus would overhear his mother and father discussing his mission as a child of destiny, Jesus found himself siding with his father, which was destined to bring heart-aches to his mother because he rejected her advice that would lead him into becoming a King like unto King David. These two attitudes were also reflective of the Jews attitude about the coming Messiah.

JESU'S STRUGGLE WITH HIS DUAL NATURE

Jesus himself had not yet become aware of his dual nature. His developing human mind was grasping with the various ideas of becoming a fisherman, a carpenter, a builder or a religious leader of a new order. While on the other hand, he was becoming aware of the calling of his spiritual mission and personality as the Creator Son of our local universe. On a few occasions, he almost doubted that he was a child of destiny, but that was slowly overcome by his absolute dedication to doing his Heavenly father's will.

Aug. 21 Jesus 12 years old.

Jesus pays more attention to music and he continues to teach the home school for his brothers and sisters.

Birth of the Lord's Prayer

Joseph, Mary, and Jesus believed all the children should be educated equally, therefore Jesus begins to home school the children. They loved his stories and he took great delight in telling them to the children. They wanted him to teach them to pray. He wanted them to learn how to have a little talk with God as a father but they would always go back to a recited, memorized prayer. The "Lord's Prayer" had its beginning during these sessions with the children. He would give them short phrases. But, like people of today, the children would recite the words without giving much thought to them. He wanted them to learn how to open their hearts and talk to God as their father. It would not be until sometime later that these lines would develop into the now known "Lord's Prayer."

Jan. Feb. Mar. Apr. May Jun. Jul. Aug. Sep. Oct. Nov. Dec.
7-A.D----------------|-----------------------|---------------------|------------------7 A.D

This was Jesus last year in school.

On Sunday night January AMOS WAS BORN Joseph & Mary's eighth child.

JESUS EXPOSED TO LIBERAL THINKING

Through his years at the synagogue, Jesus was an excellent student. He had a great advantage because he was well-versed in three languages. When Jesus finished his curses, the chazzan said he felt he had learned more from Jesus and his searching questions than he was able to teach this young lad. Many great thinkers would be invited to speak, which exposed the students to a more liberal understanding of the scriptures. When there was no passing learned man to invite to speak at the synagogue, they would ask Jesus to speak at the service.

Jesus a "Son of the commandments"

March 20th Jesus graduated from the synagogue and was turned over to his parents as an educated "son of the commandments" hereafter a responsible citizen of the commonwealth of Israel. This entitled him to attend the Passovers in Jerusalem.

Jesus first Passover

Saturday, April 4th. This year Jesus will attend his first Passover in Jerusalem. Having graduated from the synagogue school, Jesus **is qualified to proceed to Jerusalem with his parents to participate with them in the celebration of his first Passover. Mary was not going to go, because, it was not a ceremony for women. Jesus would not hear it, he strongly insisted that his mother goes with him. There were one hundred and three relatives, friends and neighbors that departed from Nazareth early Monday morning for Jerusalem. No incident in all Jesus' eventful earth career was more engaging, more humanly thrilling than this, his first remembered visit to Jerusalem. He was especially stimulated by the experience of attending the temple discussions by himself, and it long stood out in his memory as the great event of his later childhood and early youth.**

The PASSOVER FEASTFell on Saturday, April 9th, AD 7.

Jesus experience at his first Passover

As he again heard the stories about Jerusalem and the temple and all the stories of the Jewish people, recited caused the Jewish history to came alive in his heart as they traveled; standing on the Mount of Olives overlooking Jerusalem for the first time was a great thrill as his eyes soaked in the magnificent view. He could see over the top of the Golden Gate into the various courts. He could see the altar where sacrifices were laid. He could almost see into the Holy of Holies. Seeing the temple for the first time was almost more than this lad could stand. This was the House of his Heavenly Father. It was such a great delight he was beside himself and couldn't speak; a strange phenomenon for this outspoken inquisitive young lad.

A flood of spiritual illumination sweping through the mortal mind of Jesus. A messenger was sent from Salvington by Emanuel. It Said, "The time has come that you begin to be about your Father's business." Not yet 13 years, the responsibilities of a universe began to fall on this lad's shoulders. Jesus was beginning to balance his human nature and divine nature. He was a son of a man now becoming more aware of being a son of God.

All intelligent life of Nebedon, this son of God's kingdom, looked on with amazement and fascination as all this was unfolding in the thinking and actions of this young adolescent lad, a carpenter's son.

Aug. 21 Jesus 13 years old.

Jesus Becoming Aware Of His Destiny

> By the middle of February Jesus was becoming humanly aware of his mission to enlighten mankind and the revelation of God. Momentous decisions and far-reaching plans were formulating in the mind of this young youth, who to the outward appearance, was an average Jewish lad of Nazareth.

The Great Disillusionment At The Temple

The first great shock was when they were entering the temple. His mother took leave from them to go to the woman's gallery. It was the custom that only men were allowed in the temple; women were not allowed; therefore, there was a court set aside for them and their young. It never occurred to Jesus that his mother would not be allowed to accompany him to the ceremony. He was thoroughly indignant that she was made to suffer from such unjust discrimination. He let his feelings be known to his father, after which he remained quiet. But he thought deeply about it. He passed through the ceremony rituals but was disappointed by the non-caring routine manner in which they were performed. He was, however, touched by the spiritual significance of the temple ceremony and their associated worship; and they were beautiful.

Jesus refused to accept a religion that taught; God would pour out his wrath, or the anger of the Almighty onto his children. He looked appealingly into the eyes of his father and said, "My father, it cannot be true; the Father in heaven cannot so regard his erring children on earth. The heavenly Father cannot love his children less than you love me. And I well know, no matter what unwise thing I might do, you would not pour out wrath upon me nor vent anger against me. If you, my earthly father, possess such human reflections of the Divine, how much more must the heavenly Father be filled with goodness and overflowing with mercy? I refuse to believe that my Father in heaven loves me less than my father on earth."

While Joseph was taking Jesus on tour around the temple for his first time, Jesus was shocked and sickened by the spirit of irreverence of the priest and others. He received the shock of his life when he went into the court of the Gentiles. He was shocked by the noisy jargon, loud talking and cursing intermingling with the bleating of sheep, the babble of noises of the money changers, the vendors of sacrificial animals and a variety of other commercial commodities. Most of all, his sense of priority was outraged by the sight of, courtesans, the painted women, parading about within the precinct of the temple. It was a disgrace in such a sacred place. That fired up his youthful indignation, and he didn't hesitate to tell his father.

The most difficult sight, for this nature lover and animal caretaker, at the temple was when his father led him down to the priests' court beneath the rock ledge in front of the temple where the altar stood, to observe the killing of the drove of animals and the washing away of the blood from the hands of the officiating slaughter priest at the bronze fountain and the bloodstained pavement, the gory hands of the priest and the sound of the dying animals were more than this gentle lad could stand. The terrible sight sickened this boy of Nazareth; he clutched his father's arm and begged to be taken away. As they walked through the court of the Gentiles, the language and profanity were a relief from what he had just seen. Joseph led him to see the "gate beautiful" to help settle his mind, but Jesus had seen enough for his first day at the temple. Jesus did not sleep well because of the remembrances of all he had seen this first day of the seven-day event of the Passover.

Jesus Experience During The Passover

During the Passover week, Jesus kept his place with the new sons of the commandment setting outside the rail, segregating them from the full citizens of Israel. That reminded him of his youth, and therefore, he held his peace and did not ask many questions until the Passover ceremonies were over. Jesus enjoyed talking with these young sons from all the known world. Wednesday they went to Lazarus home to spend the night. Lazarus, Martha, and Bethany heard him speak, and they loved him like a brother. Lazarus was not yet 13 and could not enter the meetings. Joseph and Mary would find Jesus sitting off alone by himself in deep thought over all that had begun to flood his young mind. They had never seen him like this. They had made arrangements for Jesus to attend the best–known school. Jesus rejected all their efforts to influence his destiny that upset both parents. They just could not understand their firstborn. It can be said that Jesus sat at the feet of no man, yet he was eager to learn even from the smallest child.

Jesus Left Behind In Jerusalem

When Joseph and Mary departed, Mary assumed that Jesus was with his Father, since he was a young man now, and according to their custom, he was the responsibility of the father. Joseph, on the other hand, assumed, since Jesus was so closely attached to his mother, he would be traveling with her and the women travelers. It was not until late that evening they realized Jesus was missing. Most of that night, Joseph felt the sting of Mary's wrath. Early the next morning, they went back to Jerusalem. The following day they searched but could not find Jesus. They assumed that he was with some of their family. The next day they passed the temple. He was found in the inner chamber with the men. He astounded the rabbis, as he had done in Nazareth, with his amazing grasp of the faith and his provocative questions and responses. Mary was so angry and relieved at the same time that she marched into the men's chamber and lashed out at Jesus. Joseph always felt that the lad would be fine, but he was relieved they had found him and, Mary's wrath was now removed from him. Jesus' only response was to comely say, "I am about my father's business."

Jesus' 4th Days In The Temple

The first night Jesus stayed at Simon's house and the three youngsters were delighted to see him. The first and second day, Jesus enjoyed the quiet of the temple because most of the people had departed for their homes. When Jesus passed the brow of Olivet (an unimpressive rolling mounds that almost surrounds the city.) also called the Mount of Olives because of the olive trees that grew there and was used for oil. Jesus paused and wept over the sight. His eyes beheld a spiritual improvised people, tradition bound and living under the surveillance of the Roman legions. By the second day, he had made up his mind to take part in the discussions. He had made bold to ask the questions, most were amazed at his profound questions at such a young age. When he questions the justice of putting to death a drunken gentile who wandered outside the court of the gentile's into the forbidden and sacred presence of the temple. One of the more intolerant asked him how old he was, "Just 4 months shy of 13." He replied. He was then asked what gave him authority to be there. He replied he had completed his requirements and was approved by the synagogue in Nazareth. "We might have known; he is from Nazareth." By the third day, people were beginning to gather around to hear this young lad questioning the elders. "What really existed in the holy of holies behind the veil?" "Why should mothers in Israel be segregated from the male temple worshipers?"

Jesus' In The Temple

"If God is a father who loves his children, why all the slaughter of animals to gain divine favor, has the teaching of Moses been misunderstood?" "Since the temple is dedicated to the worship of the Father in heaven, is it consistent to permit the presence of those who engage in secular bartering and trade?" "Is the expectant messiah to become a temporal prince to sit on the throne of David, or is he to function as the light of life in the establishment of a spiritual kingdom?" He rarely replied to their responses. He continued to teach by asking questions. For four hours he continued, Simon was most astonished. The fourth day had just begun when Joseph and Mary arrived.

Jan. Feb. Mar. Apr. May Jun. Jul. Aug. Sep. Oct. Nov. Dec.
8-A.D----------------|-----------------------|---------------------|--------------8 A.D

Awareness of being a Son of God and creator of the Local Universe was beginning to weigh heavy on his mind. He was becoming more aware of his mission. He was making decisions about maintaining balance. between both natures and how to accomplish his entire mission.

The word about Jesus experience at the temple spread around Nazareth and everyone was proud of him especially his former teachers in the synagogue school.

Aug. 21 Jesus 14 years old.

The molding of Jesus' character

He is becoming a skilled worker. By Jesus' 14th birth day he had become a good yoke maker and worked well with both canvas and leather. He is also rapidly becoming an expert carpenter and cabinetmaker. He had learned a great deal about fishing and farming from his uncles. He learned about gardening and animal husbandry from his mother. His love for music and voice was inspiring. He learned masonry from his friend's father and they together learned how to make pottery and crafts. Of course, his love for drawing art and maps that got him into trouble with rabbis and the extensive conversations with travelers from around the known world was shaping this young lad's character, but none had a stronger effect than his strong ties with his loving Family.

Jesus just turned 14 years old.
Mary was 32 years old & 2 mos. Pregnant with Ruth.

A runner from Sepphoris brought the news that Joseph had been severely injured by the falling of a derrick while working on the governor's residence.

Mary takes James with her and left Jesus to care for the younger children because she did not know the extent of the injuries. Joseph died while they were on the way. They brought Joseph's body back to Nazareth and laid him to rest there.

Tuesday September 25th
Death of Joseph
The earthly father of Jesus.

With the death of his father, more responsibilities were laid on this young man's shoulders. He was now the head of the household.

James short of 9&1/2 Mariam 8 & 2 mos. Joseph 7 & ½
Simon 6 & ½, Martha 5, Jude 3, & 3 months
Amos 1 & 8 months Ruth not yet born.

Jan. Feb. Mar. Apr. May Jun. Jul Aug. Sep. Oct. Nov. Dec.
9-A.D--|----------------------------9 A.D

RESPONSIBILITIES
Jesus at the age of 14 years was not only dealing with the awareness of his son ship with God the Father and the awesome responsibilities as the creator of our local universe; he had been put into a position as the head of the household of a large family. He did everything to fill the shoes of his earthly father and be the supporter, counselor, and father to each and all of them.

RUTH

Was born Wednesday, April 17th.

Jesus and Mary both agreed that the girls should be educated just as the brothers were. He set up and ran home schooling for them as well.

Aug. 21 Jesus 15 years old.

MORE RESPONSIBILITIES
The great shock for Jesus occurred when he was 15. He went to Sapphires to receive the decision of Herod regarding the appeal taken to him in the dispute about the amount of money due to Joseph at the time of his accidental death. Jesus and Mary expected to receive a considerable amount; instead, they received a small trivial amount. They expected a large sum that would have helped support the family until they could get reestablished.

Jesus rents a considerable piece of land, divides it up and encourages competition among the family. He would work with all and was a great encouragement to the unit as a family.

JOHN ADOPTS ELIJAH'S STYLE
John (the Baptist) at the age of 16 after reading about Elijah decided to adopt his style of dress. He was more than 6 feet tall, an almost full grown a picturesque sight with his flowing hair and hairy garment and leather girdle.

JESUS PASSING FROM CHILDHOOD TO YOUNG MANHOOD
By the close of this year, they face one of the most difficult experiences of all their uphill struggles. Their hopeful courage contributed mightily to the development of strong and noble characters, in spite of the depressiveness of their poverty. Of all Jesus' Earth-life experiences the fourteenth and fifteenth years are the most crucial.
No human youth, in passing through the early confusions and adjustment problems of adolescence, ever experienced a more crucial testing than that which Jesus passed through during his transition from childhood to young manhood.

PROBLEMS AND MORE PROBLEMS
By the end of the year, Jesus could earn an equivalent to 25 cents for a long day's work. By the next year, they found it difficult to pay the civil taxes, not to mention the synagogue assessments and the temple tax of one-half shekel. During this year the tax collector tried to squeeze extra revenue out of Jesus, even threatening to take his harp.

Jan. Feb. Mar. Apr. May Jun. Jul. Aug. Sep. Oct. Nov. Dec.
10-**A.D-- --------10 A.D**

THE NEXT FEW YEARS ARE THE MOST CRITICAL YEARS OF THIS YOUNG MAN'S LIFE. HE WAS A DIVINE SPIRITUAL SON OF GOD IN THE FLESH EXPERIENCING LIFE ON THIS EVIL AND DISTRAUGHT WORLD. HE WAS TESTED MORE THAN ANY MAN ON ANY WORLD WITHIN HIS KINGDOM. HE WAS GAINING THE RIGHT TO BECOME SOVEREIGN OF HIS UNIVERSE BY EXPERIENCE.
As Jesus entered his adolescent years, he found himself the head of a large family. Within a few UNFATHOMABLE WISDOM, COMPASSION, AND LOVE for years after his father's death, most of the property was gone. During this time he was becoming more and more aware of his pre-existence at the same time he was becoming more aware of his mission in this world in the flesh for the express purpose of revealing his Paradise Father to the children of men.

Jesus began to keep his thoughts to himself.	**Simon starts school**	**Jesus is beginning to make up his mind about starting his public mission.**	Aug. 21 Jesus 16 years old.

Mary thought Jesus would become the deliverer of the Jews and it was difficult for her to understand her elder son's actions and words. But, she truly loved him and appreciated his devotion to the family.	James took charge of teaching his three sisters. It was not the norm for women to be educated nor scripturally taught but Jesus and Mary agreed that the girls were to receive instructions in both.	**Jesus was devoted to his family** This year Jesus was busy at the workbench. His superior work was admired and orders kept him busy. Any time he could he would spend with the children, telling them stories of the scriptures and they loved listening to him almost as much as he loved telling them. Jesus sold more of the property to help make ends meet. He was a real youthful thoughtful father to them all.

Jan. Feb. Mar. Apr. May Jun. Jul. Aug. Sep Oct. No Dec.
11-**A.D---------------|-------------------------|------------------------|-------------11 A.D**

Jesus had decided to enter his public life as a teacher of the truth of the heavenly Father as soon as he had reared his family and seen them married.

Jesus knew he would never become the Jewish messiah as a king and knew it was fruitless discussing it with his mother.

Jesus was a youthful lad and a good father to the children and they loved him.

Aug. 21 Jesus 17 years old.

This 17th year was a difficult year for Jesus. There was a rise of discontent as the Zealots were becoming more aggressive about not paying taxes to the Romans. As the strong nationalist party movement started to form in Nazareth they wanted Jesus to join and become an officer which would bring most of the young men with him. If he did he would have to abandon his plans of becoming a spiritual leader instead of the king his mother and so many others were expecting the Messiah to be.

Jesus was a young man and the head of a large family which required his attention and he steadfastly refused to join the uprising. By doing this he did not abandon his mission nor did he disclose his dual nature as Son of God and son of man.

As Jesus became more settled in his dual role as a son of God and a son of man it became more difficult to speak his mind with his mother and the children. They just could not understand him.

He had already decided he would not use his God given power or authority in this physical human experience. His goal was to experience all aspects of this life and let God express Himself through him. Thereby God would become visible to mankind.

Jan. Feb. Mar. Apr. May Jun. Jul. Aug. Sep Oct. Nov. Dec.
12-**A.D**--------------|-----------------------|------------------------|----------------**12 A.D**

JESUS REFUSES TO JOIN THE ZEAOLTS

even though his Mother persistently tried to induce him into joining them. Money was offered by a wealthy merchant to support his family but Jesus steadfastly refused. He insisted that the children needed a father figure who was concerned for their education and wellbeing.

Jesus was highly regarded in the city before and now some found favor with his decision and others disagreed. That divided the people in Nazareth.

This Zealot movement will later appear in Jesus' life through Judas and Barabbas.

A LOT HAPPENS THIS YEAR

Jesus takes James to the Passover. While in route they discussed their history as did he and his father before. James was a religious lad and was anxious to assume the head of the house responsibilities so Jesus could start his mission, thereafter he could join him. James wanted to hear Jesus speak in the temple, but he did not. He replied, "My hour has not yet come."

James and Simon tried to follow Jesus' example of persuasion and nonresistance, they were fairly successful. Serious family discussions were occurring more frequently.

Aug. 21 Jesus 18 years old.

Finances ease up

Financial problems eased, this 18th year when most of the property was sold. Taxes were paid, payments made for the shop and new tools were bought for the carpenter shop, the home and James began working the shop.

Amos comes down with a high fever. Death once again strikes this Nazareth family.

Dec. 3rd. Little AMOS dies

JESUS MAKES GREAT PROGRESS THIS YEAR

This year Jesus made great progress in his mind. Gradually he brought his divine and human natures together via his own decisions. With the help of only the indwelling monitor just like all other children of all post bestowal worlds.

Jan. Feb. Mar. Apr. May Jun. Jul. Aug. Sep. Oct. Nov. Dec.
13-**A.D--13 A.D**

Grieving and hardship

Work was slow, money running down, mother Mary grieving. Jesus said, "We are all doing the best we can, a mother's smile will encourage us all to do even better." Jesus infused an attitude in the family that things will get better.

THE FAMILY UNITES

Mary was getting better and began to see Jesus more as a father to her children than a son. The family was beginning to work more as a unit together. Little was said about his future mission.

Aug .21 Jesus 19 years old.

JESUS' MODE OF DISCIPLINE FOR THE FAMIELY

Jesus used a positive suggestion such as you should do this or that, rather than the negative thou shall not do this or that. That was approved and accepted by all the family especially the one that was subject to the discipline.

THE LOVE OF REBECCA

Jesus was a robust young man of Nazareth; although poor in social standing, he was not impaired. He was one of the foremost young men. Rebecca the daughter of Ezra found herself falling in love with the son of Joseph. She confided in Miriam, Jesus sister. She in tern told Mary. Mary was aroused, now I'm about to lose my son who is effectively the head of the household. What will become of the child of destiny? What more could happen?

They set out to put a stop to this before it got out of hand and before Jesus got word of it. The more they told Rebecca the more she fell in love. She thought the concern was for the loss of the head of their family and approached her father who approved the idea and would surely support their family.

Ezra invited Jesus to Rebecca's 16th birthday party. Jesus patiently and courteously listened to the father's plan, then Rebecca herself.

Jesus kindly replied that no amount of money could replace his obligations to personally rear his father's family. Ezra excused himself stating to his wife that they could never have him as a son, he is too noble for us.

JESUS' MARIAGE DECISION

Jesus thanked Rebecca for her expression of devotion adding it shall cheer and comfort him all the days of his life. He assured her that he was not free to enter into relations with any woman except with brotherly love and friendship until he had met all the obligations of rearing the family.

He added, "If I am a son of destiny, I must not assume obligations of life-long duration until such time as my destiny is made manifest." Rebecca and her father later moved from Nazareth. Although she was sought by others her reply was I am waiting for the greatest man on earth to begin his teaching of truth that I may follow him. She did and was there on many occasions even among the women at the cross.

Never again was Jesus confronted with the decision of marriage.

Jan. Feb. Mar. Apr. May Jun. Jul. Aug. Sep Oct. Nov. Dec
14.A.D----------------|-------------------------|----------------------|----------------14 A.D

JESUS VISIT WITH LAZARUS AT PASSOVER

Jesus had a strong feeling to go to Jerusalem for the Passover not knowing why. He really wanted to talk with Lazarus, and visit with Martha and Mary. Next, to his family, he loved these three most of all. His mother thought it would be good for him even though they really could not afford it.
Jesus thought a great deal as he passed on through Jerusalem stopping a few moments to look upon the temple and the gathering of throngs of visitors.
He had an increasing aversion to this Herod-built temple with its politically appointed priesthood.
Lazarus, not knowing that Jesus was coming, had already made an arrangement with friends. When Jesus suggested they celebrate Passover in Lazarus house, he replied: "we have no paschal lamb." Jesus then gave a long dissertation on the love of the Father in heaven. Then Jesus said, "let the childlike and darkened minds of my people serve their God as Moses directed; it is better that they do, but let us who have seen the light of life no longer approach our Father by the darkness of death. Let us be free in the knowledge of the truth of our Fathers eternal love."

THE 1ST PASSOVER WITHOUT THE PASCHAL LAMB

About twilight, this small group of devout Jews sat down with the unleavened bread and to celebrate Passover. Jesus called them the bread of life and the water of life.

Aug. 21 Jesus 20 years old.

JESUS BECOMING A FULL-GROWN MAN

He was born into this world as a baby of the realm, he has lived his childhood life and passed through the successive stages of youth and young manhood; rich in experience of human living, replete in the understanding of human nature.

JOSEPH GRADUATES AT THE SYNAGOGUE SCHOOL

He prepared to begin work at the home carpenter shop. Although the estate of their father was exhausted, there were prospects that they would successfully fight off poverty since three of them were now working supporting the family.

JESUS TELLS MARY

Jesus told his mother of his plans, who, after serious thought, agreed, but requested he not suggest it to the family.

JESUS ADJUSTS HIS SPIRITUAL NATURE

Jesus is slowly learning how to live the heavenly life while he continues on the earthly existence. More and more he depends on the heavenly Father. As the years pass this young man continues to experience life as it is lived in the mortal flesh.

He is becoming an expert in the divine art of revealing his Paradise Father to all ages and stages of mortal creatures. Now becoming a full-grown man of the realm-he prepares to continue his supreme mission of revealing God to man and leading man to God.

A SUMMARY OF JESUS LIFE

MICHAEL THE CREATOR SON OF GOD: was a spiritual creator son of God, called into existence from the spiritual energies of the One God by God the Father, the Only Son and the Eternal Spirit. He was given his kingdom by the Ancient of Days as described in Daniels. After finishing 6 of his 7 required bestowals he announced to his kingdom that he would bestow himself on this distraught war torn world as a man of the elemental universe. Here is where he will do battle with Lucifer, Satan, the prince of the world, the fallen angels and the angels that left their post. Here is where he will reclaim what is rightfully his.

He will take off his robes of authority and come into this world as a man. He was the only spirit born son of God, begotten, born in this world, in the flesh, as a man. He was Joshua Ben Joseph.

JOSHUA BEN JOSEPH:

Was born of the womb and passes through all phases of childhood like any other earth born child. He learned from his earthly father and mother and was led in his religious and spiritual development. He slowly began to learn his true relationship with the heavenly Father. At a young age he often sought to have a little talk with God his Father. He was guided and protected by Joseph and Mary like any other good mother and father would do. Since he had access to the entire scriptures he learned to read and study them. He was indeed a **SON OF MAN** living in Nazareth.

Aug. 21 Jesus 21 years old.

Jan. Feb. Mar. Apr May Jun. Jul. Aug. Sep. Oct. Nov. Dec.

15-**A.D**-------------|---------------------|----------------------|-----------------**15 A.D**

JESUS OF NAZARETH:

As he became a young adult he was becoming well known and revered as a bright and upcoming young man. That reputation will continue until he starts his ministry. He had already made many decisions in the process. He was tested again and again. He never failed to make the decision to do the father's will.

JESUS THE CHRIST:

Once he starts his ministry after his baptism by John he will become known as Jesus the Christ. That is a title he attained as he, the son of man, opens his heart and mind and allows the pure love of God to be expressed in him and through him.

After his death and resurrection, he becomes known as, LORD and SAVIOR. **Now he is affectingly known as Christ Michael; Creator, Reign, Master and King of his kingdom and all therein.**

Jan. Feb Mar. Apr. May Jun. Jul. Aug. Sep. Oct. Nov. Dec.
16-**A.D--16 A.D**

Jesus continues to dedicate himself to rearing the children and being the head of the house-hold. At the same time, he was formulating his plans for his earthly mission.	The family decided it would be wiser if they were not all carpenters and they diversified their trades and be able to build entire buildings.	Jesus brothers and sisters ranged in ages from 7 to 18, and he was kept busy helping them adjust themselves to the new awakenings of their intellectual and emotional lives.

This year Simon graduated from school and began working with Jacob the stone mason, Jesus old boyhood playmate and ever-ready defender.	Jesus continues to work at house finishing and cabinetwork but spent most of his time at the caravan repair shop.	**Aug. 21 Jesus 22 years old.**

At the latter part of this year, Jesus leaves James in charge of the repair shop and Joseph at the home bench.	Jesus worked 6 months with metal and acquired considerable skills at the anvil before going to Sepphoris to work with a smith.

JESUS BEGINS THE SLOW PROCESS OF WEANING HIS FAMILY

Jesus held one of his family conferences and installed James, who was 18 years of age, as acting head of the family. From this day James assumed full financial responsibility for the family. Jesus made weekly payments to his brother. Never again did Jesus take the reins out of James' hands.

Jan. Feb. Mar. Apr. May Jun. Jul. Aug. Sep. Oct. Nov. Dec.
17-**A.D**--**17 A.D**

FINANCES BEGIN TO STABELIZE

3 were working; Miriam earns income from the sale of milk & butter. Martha had become an expert weaver. The shop was now over 1/3 paid for.

JESUS TAKES SIMON TO PASSOVER

They took routes that covered most of Palestine north of Jerusalem. They took three weeks, it was the freest time Jesus has had since the death of his father.

Simon's Passover

Simon greatly enjoyed his visit to Jerusalem. He was duly received into the commonwealth of Israel at the Passover consecration of the new sons of the commandments.
Simon never forgot what Jesus taught him on this trip. He had always loved Jesus, but now he felt that he had begun to know his father-brother. They had many heart-to-heart talks. They arrived home Thursday noon, and Simon kept the family up late that night.

JESUS MEETS YOUNG HELLENIST

While in Jerusalem Jesus met a young Hellenist with whom he had long conversations with. It impressed this young man so much that he began to teach what he had learned and some 15 years later became a believer in Jesus. He had no inclination that it was the same person. He was stoned to death because of his boldness and became the first martyr for Christianity.

Aug. 21 Jesus 23 years old.

JESUS SPENDS TIME IN DAMASCUS

Jesus Spent 4 months as a guest of a wealthy merchant he had met in Jerusalem. This man was part Jewish and he wanted to establish a school of religion that would out rival Alexandria. He proposed that Jesus head the school, which was one of the greatest real temptations for Jesus in his purely human experience. He helped them lay out their plans giving advice and directions, but he refused to accept the offer.
Jesus became acutely aware that many would try to formulate a religion about him instead, which might become competitors with the gospel of the kingdom. He did not want men to believe in him unless their hearts were responsive to the spiritual realities revealed in his teachings.

Jan. Feb Mar. Apr. May Jun. Jul Aug. Sep. Oct. Nov. Dec.
18-**A.D--------------|------------------------|-----------------------|----------------18 A.D**

JESUS' FIRST YEAR OF RELATIVE FREDOM FROM FAMILY RESPONSIBILITIES

James was very successful in managing the home with Jesus helping counsel and financially.

MORE LOVE & MARRIAGE IN THE AIR

Miriam approached her brother-father with her plans. Jacob, the younger stone mason, onetime self-appointed champion of Jesus, now a business associate of James and Joseph, had long sought to gain Miriam's hand in marriage. After Miriam had laid her plans before Jesus, he directed that Jacob should come to him making a formal request for her and promised his blessing for the marriage just as soon as she felt that Martha was competent to assume her duties as the eldest daughter.

LOVE IS IN THE AIR

James was in love and wanted to get married. He pointed out that Joseph would be 18 soon. Jesus would consent if after 2 years, and Joseph was fully trained to assume the direction of the home.

Aug. 21 Jesus 24 years old.

JESUS AGAIN DECLINES AN OFFER TO BECOME INVOLVED IN ORGANIZED RELOGION

Jesus meets with five prominent Jews of Alexandria. They offered to establish him in their city as a religious teacher, offering to begin with, the position of assistant to the Chazan (Cantor) in their chief synagogue. Jesus listened to all they had to say, thanked them for their confidence and, in declining to go to Alexandria, in substance said, "My hour has not yet come." They presented him with a purse in token of the esteem of his Alexandrian friends and in compensation for the time and expense of coming over to Caesarea to confer with them. But he likewise refused the money, saying: "The house of Joseph has never received alms, and we cannot eat another's bread as long as I have strong arms and my brothers can labor."

The remainder of this year was the most uneventful six months of his whole career. He enjoyed this temporary respite from the usual program of problems to solve and difficulties to surmount. He communed much with his Father in heaven and made tremendous progress in the mastery of his human mind. He continued to teach the evening school three times a week. He read the Scriptures often in the synagogue on the Sabbath, He visited with his mother, taught the children, and in general conducted himself as a worthy and respected citizen of Nazareth in the commonwealth of Israel.

Jan. Feb. Mar. Apr. May Jun. Jul. Aug. Sep. Oct Nov. Dec.
19-**A.D-------------|--------------------|---------------------|------------19 A.D**

JESUS TAKES JUDE TO PASSOVER / JUDE ARRESTED

Jesus feared trouble if he took his young brother through Samaria. Already at Nazareth Jude had got into slight trouble several times because of his hasty disposition, coupled with his strong patriotic sentiments. They arrived at Jerusalem and were on their way for a first visit to the temple, the very sight of which had stirred and thrilled Jude to the very depths of his soul when they chanced to meet Lazarus of Bethany. While Jesus talked with Lazarus, a Roman guard made some improper remarks regarding a Jewish girl who was passing. Jude flushed with fiery indignation and was not slow in expressing his resentment of such an impropriety directly to, and within hearing of the soldier.

The Roman legionnaires were sensitive to anything bordering on Jewish disrespect; so the guard promptly placed Jude under arrest. This was too much for the young patriot; he delivered a voluble denunciation of pent-up anti-Roman feelings, all of which only made a bad matter worse. Jude, with Jesus by his side, was taken at once to the military prison. It took two days, which caused him to miss the Passover.

BACK HOME AGAIN

Jude's out spoken manner caused him to miss the ceremony, but, he will go to Passover the following year. Jesus and Jude had a long discussion on their way home. Jude began to understand his brother father more. Jesus said nothing of the problems when they returned. Jude did tell the family everything.

Aug. 21 Jesus 25 years old.

LIFE GETS BACK TO NORMAL FOR A WHILE

The family becomes more involved in their daily affairs and Mary almost forgot that he was to be a child of destiny.

Jan. Feb. Mar. Apr. May Jun. Jul. Aug. Sep. Oct. Nov. Dec.
20-**A.D**--**20 A.D**

Jesus Slowly Removes Himself From Family,

But He Never Forgot The Children

This year his seasons of deep meditation were often broken into by Ruth and her playmates. Always was Jesus ready to postpone the contemplation of his future work for the world and the universe that he might share in the childish joy and youthful gladness of these youngsters, who never tired of listening to Jesus relate the experiences of his various trips to Jerusalem. They also greatly enjoyed his stories about animals and nature. The children were always welcome at the repair shop. Jesus provided sand, blocks, and stones by the side of the shop, these youngsters flocked there to amuse themselves. When they tired of their play, the more intrepid ones would peek into the shop, and if its keeper were not busy, they would make bold to go in and say, "Uncle Joshua, come out and tell us a big story." Then they would lead him out by tugging at his hands until he was seated on the favorite rock by the corner of the shop, with the children on the ground in a semicircle before him. How the little folks did enjoy their Uncle Joshua. They were learning to laugh and to laugh heartily. It was customary for one or two of the smallest of the children to climb upon his knees and sit there, looking up in wonderment at his expressive features as he told his stories. The children loved Jesus, and Jesus loved the children.

The family was marrying and settling down, even Jude who became a fisherman. Jesus spent much time with each of the family. They were all becoming aware that Jesus would soon be leaving the family.

Jesus 26 years old.

JESUS BECOMES FULLY AWARE OF WHO HE WAS AND HAD ESTABLISHED HIS PLANS

Jesus of Nazareth became strongly conscious that he possessed a wide range of potential power. But he was likewise fully persuaded that this power was not to be employed by his personality as the Son of Man, at least not until his hour should come. At this time he thought much but said little about the relation of himself to his Father in heaven. And the conclusion of all this thinking was expressed once in his prayer on the hilltop when he said: "Regardless of who I am and what power I may or may not wield, I always have been, and always will be, subject to the will of my Paradise Father." And yet, as this man walked about Nazareth to and from his work, it was literally true — as concerned a vast universe — that "in him were hidden all the treasures of wisdom and knowledge."

James was given complete authority over the home, and Jesus promised to send funds to help. At last Jesus' brothers had chosen, and were established in, their lifework. The stage was being set for Jesus' departure from home.
And thus did Jesus make ready to enter upon the second and home-detached phase of his adult life before the public entrance upon his Father's business.

Jan. Feb. Mar. Apr. May Jun. Jul. Aug. Sep. Oct. Nov. Dec.
21A.D-----------------|----------------------|----------------------|---------------21A.D

JESUS AND ZEBEDEE BUILD BOATS

Jesus went to Tiberias, and finding little to interest him; he passed on successively through Magdala and Bethsaida to Capernaum, where he stopped to pay a visit to his father's friend Zebedee. Zebedee's sons were fishermen; he himself was a boat builder. Jesus of Nazareth was an expert in both designing and building, Zebedee had long known of the skill of the Nazareth craftsman. Zebedee had contemplated making improved boats; he now laid his plans before Jesus and invited the visiting carpenter to join him in the enterprise, and Jesus readily consented.

Jesus worked with Zebedee only a little more than one year, but during that time he created a new style of boat and established entirely new methods of boat making. Jesus had long worked alone in the world, that is, without a father, and greatly enjoyed this period of working with a father-partner.

Aug. 21 Jesus 27 years old.

Zebedee's wife, Salome a relative of Anna's, greatly admired Jesus. She loved him as she loved her own sons, James, John, and David. Capernaum was a strong Roman military post, and the garrison's commanding officer was a gentile believer in Yahweh, and he took it upon himself to build a beautiful synagogue in Capernaum. Jesus conducted the services in this new synagogue more than half the time this year.

At the synagogue, Jesus found many new books in the library chests, and he spent at least five evenings a week in intense study. One evening he devoted to social life with the older folks, and one evening he spent with the young people. There was something gracious and inspiring about the personality of Jesus which invariably attracted young people. He always made them feel at ease in his presence. Jesus conducted a conference of questions and answers each evening after supper before he departed for the synagogue to study. The youthful neighbors also came in frequently to attend these after-supper meetings. To these little gatherings, Jesus gave varied and advanced instruction, just as advanced as they could comprehend. He talked quite freely with them, expressing his ideas and ideals about politics, sociology, science, and philosophy, but never presumed to speak with authoritative finality except when discussing religion — the relation of man to God.

Once a week Jesus held a meeting with the entire household, shop, and shore helpers, for Zebedee had many employees. And it was among these workers that Jesus was first called "the Master."

This year Jesus made great advances in the ascendant mastery of his human mind and attained new and high levels of conscious contact with his indwelling Thought Adjuster. This was the last year of his settled life. Never again did Jesus spend a whole year in one place or at one undertaking. The days of his earth pilgrimages were rapidly approaching.

All this year Jesus sent money each month to James.

Jan. Feb. Mar. Apr. May Jun. Jul. Aug. Sep. Oct. Nov. Dec.
22A.D------------------|-----------------------|-----------------------|---------------22A.D

JESUS PREPAIRS TO LEAVE HIS FRIEND ZEBEDEE

Jesus only worked a little over a year as a boat designer and builder with Zebedee. The business did so well a large sum of money was built up and entrusted to Zebedee. Jesus only used a small amount as needed. In March A.D. 22 Jesus takes leave of Zebedee and Capernaum. He asks for a small sum of money to defray the cost of his travels, and ask that money would be sent monthly to his family. John Zebedee agreed and promised to stand by Jesus mother and family as they were his own, saying, "Go about your business my teacher, do your work in the world; I will act for you in this matter."

A PROMISE FROM JOHN ZEBEEDE.

Jesus told John Zebedee of his plans to travel throughout the Mediterranean. He enjoined him to tell no one, not even his own family. Zebedee was true to his word and told no one of Jesus' whereabouts.

Before leaving Jesus enjoyed celebrating Passover with all his friends as one big happy family.

April 26 Jesus began his travel in earnest. He traveled throughout the Roman Empire and beyond. He met many people with whom he shared his wisdom and knowledge. Some of whom would hear his teachings later without realizing that they had known this man before.

Aug. 21 Jesus 28 years old.

HEARTBREAK AND HOPE

Jesus's family had heard nothing and had almost given up that Jesus was even alive. Each month Zebedee and his son John would visit with Mary and the family bringing gifts as they promised Jesus. Those visits warmed Mary's heart and kept hope alive.

During this time the Nazareth family was getting along very well. Their family income had increased. They were marrying and the family was increasing and becoming self-sufficient.

Jan. Feb. Mar. Apr. May Jun. Jul. Aug. Sep. Oct. Nov. Dec.
23-**A.D**------------|---------------------|--------------------|-------------**23 A.D**

Extended travel throughout the Mediterranean

Jesus met a wealthy merchant Gonad and his son, Ganid from India. Jesus agreed to travel and tutor the son who was a spiritual seeking lad about 17 years of age and was an eager and apt learner.

Jesus did not return until December A.D. 23. Jesus was learning about the earthly experience of mankind and planting the seeds of God's love in all those he met.

On these trips, Jesus met hundreds of humankind. He met and loved all manner of men: rich and poor, high and low, black and white, educated and uneducated, cultured and uncultured, animalistic and spiritual, religious and irreligious, moral and immoral. He was coming to know mankind with his experiences. All entities within his kingdom would be able to see and share those experiences.

He would discuss almost any topic the people were led to ask: social events, geographic, political, tribal affairs, religious or morality he was always ready to engage the people he met.

Aug. 21 Jesus 29 years old.

Jesus never hesitated to speak with peoples of different beliefs always directing the thoughts toward the One God. He would say, the will of God is the way of God, to do the will of God, therefor, is the progressive experience of becoming like God. God is the source of and destiny of all that is good and beautiful and true. Will is the deliberate choice of self-conscious being making decisions based on intelligent reflection.

Jesus's travel on this tripInclude many srories:

from Jerusalem to Jappa-then Caesarea-Alexanderia-Crete-Cathage-Malta-Syracuse-Messin-Naples &Capua-Rome-Northern Italian lakes-Nichopolis-Corinth-Athens-Ephesus-Rhodes-Cyprus-Antioch-Sidon-Damascus-Babylon-Ur-Susa-Charax-Babylon-Damascus-Bethsaida. This was a long trip one ot which Jesus enjoyed emensley.

NOTE **This web site has a multitude of maps for those who are interested in learning more about Jesus travel:** http://www.urantia.org/in-his-steps

Jan.	Feb.	Mar.	Apr.	May	Jun.	Jul.	Aug.	Sep.	Oct.	Nov.	Dec.

24-A.D--------------|----------------------|-----------------------|--------------24 A.D

The family thought Jesus was studying somewhere. Simon and Jude were waiting for his return to get his blessings for their wedding. Early March they had a double wedding. Only Ruth was still with Mother Mary. Jesus suggested they move into a house in Capernaum where they stayed for the rest of Mary's life.	**On this trip Jesus, as always, he was willing to discuss everything with all who would listen. He always took the discussions to the absolute sovereignty of God and doing the will of God was supreme. He pointed out that there could be no peace on this world until all mankind developed a world brotherhood with tolerance, respect, and love for one another. If world politics, religions and societies could develop a standard, or laws, by which all could and would comply with that same brotherly love and respect without trying to push their individual beliefs on others. There could be peace on earth as was willed by God. As long as mankind tries to force dictate or otherwise subdue mankind's will there would always be wars and rumors of wars.**

World law must come from and be enforced by world governments. Collective security will never afford peace until the collectivity includes all mankind. The spiritual brotherhood of man will forever insure good will among all men. There is no other way.

The leader of a large caravan got sick and Jesus volunteered to stand in for him. April A.D. 24 they departed Nazareth –Damascus-through Assyria-to Lake Urmia-then though Media and Parthia on to the Caspian seashore. On This trip He was experiencing, first hand, more and more about the people of this world.

Jesus Takes One Year As A Caravan Leader

From Damascus To Caspian Seashore

Aug. 21 Jesus 30 years old.

Jan. Feb. Mar. Apr. May Jun. Jul. Aug. Sep. Oct. Nov. Dec.
25-**A.D**--------------|----------------------|------------------------|---------------**25 A.D**

JESUS TAKES ONE LAST SOLO JOURNEY

Jesus knew his world travel was nearly over. He took one last trip which lasted about 4 months. It was a solitary visit through Syria, Capernaum, Nazareth, Tyre, Sidon Antioch, Caesarea, Jappa, Jamnia, Ashdod, Gaza, Beersheba, Palestine, Beersheba, Dan, Hegron, Bethlehem, and many other cities. This was a solo trip for Jesus and the people knew him by many different titles. Jesus was coming to know mankind with all its frailties. At the same time, he was gaining absolute control of his own human mind and attuning it to the inner spiritual self.

Aug. 21 Jesus 31 years old.

JESUS IS LEAD TO MOUNT HERMON

The Indwelling Thought Adjuster led Jesus to Mount Hermon away from the dwelling places of humans so he could culminate his earthly mind and concentrate on the remainder of his life work on the earth.

JESUS MAKES READY FOR THE MOUNT HERMON TRIALS AND TEMPTATIONS

In the middle of August, In the foot hills of Mount Hermon Jesus made ready his supplies, purchased a donkey and hired a lad named Tiglath to bring him supplies up to a place, about 6,000 feet above sea level, where they prepared a place where Tiglath would place new supplies for Jesus twice a week.

After traveling a short distance further up the mountain Jesus stopped to pray and asks that the guardian seraphim depart him and stay with Tiglath. He asked that he be allowed to meet this last struggle alone. It was granted. He went into the great test with only his indwelling Adjuster to guide and sustain him. Jesus ate frugally only fasting one or two days at a time while on Mount Hermon.

6 WEEKS

For the last 3 weeks of August and the 1st. 3 weeks of September Jesus was on Mount Hermon. During these 6 weeks, he completed his moral task, achieving the circle of mind –understanding and personality control.

THE TEMPTERS

The superhuman beings that confronted him, and with whom he wrestled with in spirit and with whom he defeated were real. They were his archenemies in the system of Satania.

JESUS RESPONSE

Jesus requested to meet the tempters as the son of man, Joshua Ben Joseph a human without assistance from the heavenly host. To the many proposals and counter proposals of the emissaries of Lucifer. Jesus only made reply: "May the will of my Paradise Father prevail, and you, my rebellious son, may the Ancient of Days judge you divinely. I am your Creator Father; I can hardly judge you justly, and my mercy you have already spurned. I commit you to the adjudication of the judge of a greater universe. Jesus had won the unquestioned sovereignty of his universe

a time to wait

When Jesus departed Mount Herman he gave Tiglath the donkey and returned to see the family and on to the house of Zebedee where he worked and visited nearby cities. Jesus was noticeably changed. He was quieter and more contemplative than before. He heard the stories of John's ministry, saying "The kingdom of heaven is at hand; repent and be baptized."

Jan. Feb. Mar. Apr. May Jun. Jul. Aug. Sep. Oct. Nov. Dec.
26-A.D--------------|----------------------|----------------------|--------------26 A.D

JOHN THE BAPTIST

When the time came Zachariah and Elizabeth took John to the Nazarite brotherhood near the Dead Sea to be inducted into this order for life. He took the vow to never cut his hair, to drink no intoxicating drinks and never touch the dead. Like Sampson and the prophet Samuel. They were held with great respect similar to the High Priest. Nazarites for life were the only ones, other than the High Priest, allowed to enter the Holy of Holies in the temple.
John considered himself to be the last of the old prophets and the first of the new.
By the time Zachariah and Elizabeth's death, John had deduced that the messiah would have to be one that sat at the right hand of God. Therefore, his kingdom would haft to be a spiritual kingdom thus he preached "The kingdom of Heaven is at hand".

Aug. 21 Jesus 32 years old.

THE SON OF GOD
THE SON OF MAN

Joshua Ben Joseph had grown from an astute young lad to a wise and thoughtful young man. Who now gained complete control and of his mind and soul becoming one with the spirit with complete consciousness of who he was, and what his mission was to be. He had made monumental decisions not to use his God given power and authority to accomplish his mission. He would use them to help others, not himself. He was a Son of God, but he would be faithful and true as a son of man.

IT'S TIME

When John reached a point near Pella, Jesus laid down his apron and tools and said, "My hour has come."
He went and presented himself to John to be baptized. There was a large crowd. John did not see Jesus until he was standing in front of him. When John recognized Jesus the ceremony was halted while he greeted his earthly cousin. He asks Jesus why he came down into the water to meet him. Jesus replied, "To be subject to your baptism." John replied, "But it is I who needs to be baptized by you." Jesus replied, "Bear with me now, for it becomes us to set this example for my brothers standing here with me, and that the people may know that my hour has come."
At noon on Monday, January 14 A.D. 26 John the Baptist did baptize Jesus and his two brothers James and Jude. After they were baptized John dismissed the crowd assuring them that he would begin tomorrow at noon. As the crowd was leaving the four men still standing in the water heard a strange sound and appearance over Jesus head the voice said, THIS IS MY BELOVED SON IN WHOM I AM WELL PLEASED."

40 DAYS IN THE HILLS

A great change came over the countenance of Jesus, and coming up out of the water in silence he took leave of them going toward the hills in the east. No man saw Jesus again for 40 days.
This was not the time of great temptations of Jesus that occurred on Mount Herod. This was a time of great planning for his ministry.
When John saw Jesus approaching from the hill John stood up on a large rock and proclaimed: "Behold the Son of God, the deliverer of the world! --- I baptized with water He shall baptize with the Holy Spirit."

Jesus' Ministry

Jesus started his ministry at the peak of Johns ministry. There was a great difference between John the Baptist and Jesus. John was an eager and earnest worker driven by the spirit; Jesus was a calm and happy worker. John's message was Repent and escape the wrath to come. His message was not to console the people but to condemn them by bringing to light their shortcomings.

John spoke of the Kingdom of heaven, but he never entered therein. Jesus entered into the kingdom of heaven and then **invited** all others to Repent and join him. It was always followed by the Gospel, the good tidings of the joy and liberty of the new kingdom.

The primary principals upon which Jesus built most of his teachings were rather simple, but it was difficult for the people and their consciousness of that time. 1st. There is only One God. Jesus did not expound on this very much with the Jewish people; it was understood by them and other descendants of Abraham. 2nd. God created all creation from His Omni-powerful force and energies. He needed no other. 3rd. Since God created all creation, it came from His energy He, therefore, is the Father of all His creation. 4th. He is the Father of all, He, therefore, is our Father and we are His sons and daughters. 5th. that then makes us brothers and sisters. 6th. that leads to the Fatherhood of God, and the brotherhood and sisterhood of all humanity. 7th. That would hold true to all levels, spheres, and dimensions of God's creation. There is no Fatherhood without Sonship, there is no sonship without fatherhood, and there is no brotherhood without sonship.

The second principle was God is a God of love, compassion, and mercy, not a God of wrath, vengeance, and jealousy. The evil ones had set up erroneous rituals misleading humanity. They taught humanity that God was sitting on His throne somewhere far above the clouds and could not see us and did not care about us, only they, the evil ones, cared. They taught that God had His angels running around spying on humanity and reporting back to Him. God would then send down His blessings or rod of discipline on the heads of humanity. They taught it was necessary for humanity to appease them, the evil ones, to gain their approval and their blessings. For humanity to protect himself, he thought it was necessary to make sacrifices or barter with God to win His favor.

Jesus knew God was a God of love, **unconditional love**, and was approachable to all His children, not just a few. We had but to ask with a sincere heart and we, Mankind, would receive His blessings. God is winning us to Him with His love for us. God loves us as we are and does not need any sacrifice nor ritual of any kind. In fact, humanity has nothing of value to offer God except his free will (which is a sacred gift from God). It must be freely given, as God gave it, motived by love, unconditional love.

Jesus not only told us these things, but he also showed us: Jesus opened his heart and allowed the will, wisdom and love of God the Father, the Only Son, and the Eternal Spirit to be expressed in him and through him. Therefore, we can see not only Jesus, but we can see them all, through Jesus's actions and words.

The Two Gospels Of Good News

In The New Testament

The 1st.

It was taught by Jesus. It was the fatherhood of God and the brotherhood of mankind. He always taught to hold God first and foremost in our heart and minds, actions and deeds and love everyone else as our brothers and sisters because we are that in the family of God. Unfortunately, that was looked at as the brotherhood of just one religion or faith and was not to include all others. There is only one family of God and it includes every man, woman, and child of every race, color and creed on every level, sphere and dimension of God's creation.

The 2nd.

This was the gospel **about Jesus the man**. It was started by Peter after the day of Pentecost, the bestowal of the spirit of truth, and was fortified by the other decuples even though Jesus had forewarned them not to do that. It is easy to understand the exuberance of the disciples because they had seen, touched, talked with, and supped with Jesus the master after his physical death, and before his final ascension back to the Father's house.

A lot of the truths that Jesus taught was carried along with this new gospel about Jesus. It does, however, put that old fatal stumbling block of (my way or the highway,) in the path for those of other faiths. How unfortunate for mankind.

Two World Orders

THE 1ST.

This will be mankind's attempt to unite the world under a one world government: with one world religion one type of money and one marketing process for all mankind.
It would give controls that could have a lot of benefits for mankind by reducing fraud, greed, imbalances in health, education, welfare etc.
If each nation would give up its desire for their own sovereignty to a representative World government in order to remove the threat of wars and rumors of wars. All peoples would have to develop a respect for others of a different faith as well as for those of no faith.
Mankind would have to make a lot of changes. Unfortunately, mankind is so bogged down in their own selfish ways and will not be willing to give up anything, because of their lack of trust.
All of this will probably lead to that which is prophesied in the scriptures as
THE END TIMES.

THE 2ND .

This will be that period called moving into the light. When mankind will be exhausted trying to fend for themselves. It will appear that the evil will win the battle. Mankind will have undergone the spiritual, mental and physical attitude adjustments and will be ready to be aided by the Love of Christ and the angels of God.
Victory will be had and all will know that it was the One God that won the war over the evil forces. Christ will reclaim what is rightfully his.
The Christ will return and establish His world order. It will be built upon the brotherhood of mankind and a world government of mankind. World Laws will come into being for the betterment of mankind and must be agreed to by all nations and enforced by the world government - the sovereignty of all mankind. National affairs will be handled by national governments.
Local government will be handled by local governments.
The political sovereignty of representative mankind will bring lasting peace on earth and the spiritual brotherhood of man

Where We Are At Now

This world's nations have not possessed real sovereignty. They have never had a sovereignty which could protect them from the ravages and devastations of world wars. In the creation of the global government of mankind, the nations are not giving up sovereignty so much as they are actually creating a real, bona fide, and lasting world sovereignty which will henceforth be fully able to protect them from all war. Local affairs will be handled by local governments; national affairs, by national governments; international affairs will be administered by the global government.

World peace cannot be maintained by treaties, diplomacy, foreign policies, alliances, balances of power, or any other type of makeshift juggling with the sovereignties of nationalism. World law must come into being and must be enforced by a world government, the sovereignty of all mankind. The individual will enjoy far more liberty under a world government. Today, the citizens of the great powers are taxed, regulated, and controlled almost oppressively, and much of this present interference with individual liberties will vanish when the national governments are willing to truste their sovereignty as regards international affairs into the hands of a global government. Under global government, the national groups will be afforded a real opportunity to realize and enjoy the personal liberties of genuine democracy. The fallacy of self-determination will be ended. With global regulation of money and trade will come the new era of world-wide peace. Soon a global language may evolve, and there will be at least some hope of sometime having a global religion or religions with a global viewpoint.

Collective security will never afford peace until the collectivity includes all mankind. The political sovereignty of representative mankind government will bring lasting peace on earth, and the spiritual brotherhood of man will forever insure good will among all men. And there is no other way whereby peace on earth and good will among men can be realized.

World Order For The Brotherhood Mankind

The sovereignty of the ONE GOD

The ONE GOD conceived, planned, and created all creation. He is the force and power from which all creation came and in whom all creation exists. He has established an order and procedures for all creation. He and He alone has complete authority with absolute sovereignty which extends out and down throughout all of His creation. Our world is but one world among so many. We are not alone; God has delegated authority down throughout all creation to our world. He, through the chain of authority, has assigned our world with a spiritual prince of this world. Unfortunately, he joined with Lucifer, and Satan in the war that took place in heaven and brought havoc down onto our world. It has since been governed vicegerent by Machiventa Melchizedek who is High Priest of the most high God.

Sovereignty Of The Individuals Will

God loves each and all of His children with divine unconditional love. In his great wisdom and understanding, He has given all His children FREE WILL. Via that free will we have the divine right to choose to follow God and His plan, or choose to follow our ways which more often than not leads to our demise. God does not love nor show favor to any one person, group, sex nor religious order or faith. He unconditionally loves us all. He knows our strengths and our shortcomings He knows all the hardships we as individuals and as the world have endured, better than we. With infinite and divine love, He wants only that which is best for us, but the choice is ours.

Sovereignty Of Collective Humanity

This world is but one of an untold number of inhabitable worlds in all creation. We human beings were given complete authority over all of this world. All of which fall under the spiritual authority of God's divine organization. Therefore we fall under the spiritual authority of the acting Prince of this world Melchizedek. Our world is truly blessed to have him as our prince. Our world has also been blessed by the Great Prince Michael, creator, lord, and master of our Local Universe, our galaxy. Almost all eyes are on us, this world. Because of the war in heaven: Lucifer, Satan, the prince of this world, the fallen angels and the angels that left their post have tried to take our free will from us, but because of the divine love and awesome efforts of so many of God's spiritual sons we now have an opportunity to join with them and raise this world out of the nightmare we have endured.

It is within the power of the collective consciousness of the good peace-loving brothers and sisters of this world, to stand up and demand our leaders to quit their warring ways and develop a <u>Global Order for the Brotherhood of Mankind</u> whose primary purpose is to establish and maintain law and order over the air, land and waters of this world. To eliminate all warmongers who would attempt to take over others. Each country would relinquish the authority to declare and make war on their behalf that would be the responsibility of the sovereignty of the brotherhood of humanity. Each country would still have the responsibility of policing and maintaining its own social, economic, immigration laws, ensuring that all their laws comply with the brotherhood of all humanity. Freedom is the gift of civilization made possible by the enforcement of LAW.

Religion makes it spiritually possible to realize the brotherhood of men, but it will require a humanity government to regulate the social, economic, and political problems associated with such a goal of human happiness and efficiency.

A layout for World Order

for the Brotherhood of Mankind.

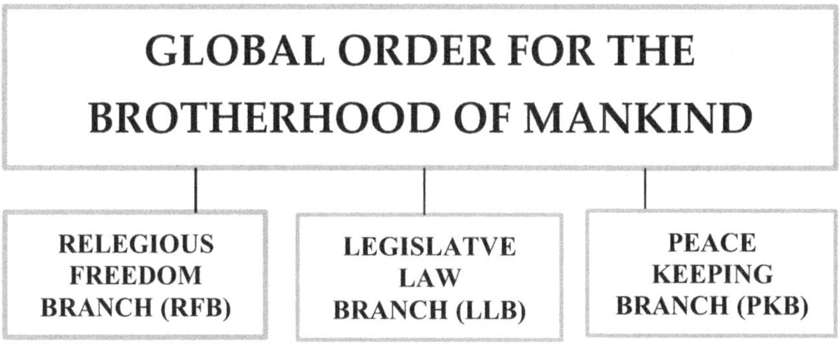

The Global Order for the Brotherhood of Mankind **(GOBM)** Should be headed by a triune one from each of the branches elected from within its appointees. They are to speak with one voice through a rotating spokesperson. Each of the triune is to be replaced one at a time after finishing their appointed time.

The Religious Freedom Branch **(RFB)**

Should be one or two appointees from each official religious orders. Their sole purpose should be to find ways to ensure the freedom of religions for all induvial's of this world. Free will is a precious and divine gift from God and should not be trampled on nor abused by anyone. That freedom of individual will to believe or not believe is sacred. RFB responsibility should hold those free choices sacred for all. Therefore, their mandates should always reflect and protect them.

The Legislative Law Branch **(LLB)**

Should be one or two experts on law appointed from each country. Their primary purpose is to create laws designed to enhance the freedoms and will for the Brotherhood of humanity without favor to anyone or groups of countries. All laws must take into consideration the freedom of religion and rights for each and all of humanity.

The Peace Keeping Branch **(PKB)**

Should have the sole responsibility of maintaining peace on the land, at sea, and in the air for all countries without favor for anyone or group of countries. All are equal and are to be treated with respect. If a dispute arises between countries, it is to be brought to the **GOBM** to find a resolution via feedback from the three branches. Their findings are to be final and fair for all with the greatest respect for the individuals and all humanity.

This unified military force will be contributed to by all countries on an equal basis for men, machine, and money. Once a decision is made to use force by the **GOBM** the responsibility for planning and carrying out will be the sole responsibility of the **PKB** and must stay within the parameters agreed to by all three branches.

The approach to finding solutions should be from the appropriate branch or branches calling together all parties involved, and with the greatest respect, they should search for a peaceable solution. In the event the dispute is between nations, there should be a mandatory meeting called between the leaders of those nations. It must be monitored by the GOBM and at least two representatives from each of the three branches trying to reach a peaceable solution for all involved. If that fails and the suggestions are refused, then and only then is it to be turned over to the PKB for determining the amount of force to be used to bring order. In this manner, all nations should learn to live in harmony and peace.

Conclusion

The one monolithic force or power from which all creation came, is alive and is with consciousness. It is that one we call the One God. All creation came from the Omni-power of that one force it is the string that holds all creation together. Nothing was created from nothing; it was created from the power from within the One God. The One God is a multi-dimensional Creator, and He has created a multi-dimensional creation. God planned out all creation with great wisdom and understanding within His mind and left nothing to chance, but He did leave a lot to choice for those whom He gives the precious gift of free will. The One God projected forth a great amount of His power and energy into that which has become known as Paradise or Isle of Paradise. Therein He projected His willpower into that power and began to arrange it as he desired. His WILL is that which we call GOD THE FATHER. Then He turned and called forth another like unto Himself, having the same total essence from within the One Supreme Power and is His absolute pure unconditional love. That love is the ONLY SON of God. The wisdom and understanding of the One manifested forth from the One and is called THE ETERNAL SPIRIT. (A special note: the Mother Spirit and the Holy Spirit are exact copies of the Eternal Spirit. The different names only signify what part of God's creation wherein they function.) Together the three (God the Father, God the Son and God the eternal Spirit) are commonly referred to as the Eternal Triune or Holy Triune.

The Isle of Paradise is the first heaven referred to in the King James Bible, Genesis 1:1 "In the beginning God created the Heaven---" it is singular not plural. It is the permanent abode of God the Father. Paradis, in all of its glory, is the throne of God. In God's great plan of creation, He had a plan for the organization and function for every aspect and dimension. The Father and the Son held fast in their mind every quality and attribute that would be needed by every spiritual personality to create and establish absolute control of every aspect and dimension on every level and plain of God's creation. The Eternal Spirit moved within them and brought forth ALL the heavenly host; they were called the Light in Genesis 1:3.The Only Son and the Eternal Spirit conjointly created the Central Heaven. (the second heaven) It was completed, complete and in perfect harmony with Paradise and had reign over all else that was created.

God the Father and the Only Son projected out a massive amount of spiritual energy force, around Paradise and the Central Heaven, into which they cast down the seven thrones which become the seven super universes within the third Heaven. They are the seven super universes. The Ancient of Days assumed their positions of authority. (There are three Ancient of Days in each Super Universe). The Only Son and the Eternal Spirit sent forth from that Light Powers, Principalities, and the Angelic Host. Their assignment was to create, maintain, and control a variety of Local Universes which are Galaxies. A Great Prince of the order of Michael who is like unto a Son of the Only Son and a Grandson to God the Father. And a perfect copy of the Eternal Spirit was brought forth (which is his conjoint worker) called the Mother Spirit; her actions are called the Holy Spirit. Their task is to create a Local Universe wherein new life can be born and given the opportunity to CHOOSE to join with them and God the Father's great plan of creation. As God, the Father is the Father of all creation; this great prince is the father of all his creations. Therefore he is our creator father. We like unto them are the father (parents) of our children. It then becomes easy to understand; God the Father is the Father of all creation. The Only Son is the conjoint father of all creation, and the Great Prince is our creator father, and we are the father of our children. The task and function of all of them, and we are to rear our children and teach them by example about the unconditional love of God.

Everything was going well with only small disturbances here and there in our Great Prince's kingdom. About the time our world was forming and ready to become inhabited, Lucifer, a son of the morning, who was assigned as our local system began to entertain thoughts of self: self-aggrandizement, self-glorification, and self-gratification which became like a virus and began to occupy much of his thoughts. That happened about 200,000 years ago. He was to be in charge of 1,000 inhabitable worlds, but at that time there was only about 607 which were developing. That virus grew until he convinced his Liaisons Officer, Satan to join his erroneous thoughts and teachings. The now growing cancer infected about 37 worlds. Our world was one of them. In fact, it became the most distraught and engrossed in that rebellion. Humanity was the victims, misled by those evil forces. That war was over in heaven when Gabriel, the Morning Star, did battle with Lucifer. That war was over in this world when the Great Prince Michael the spiritual Creator Son of God who is Lord of our local universe, our galaxy, took off his robes of authority and came to this world as a son of man. As a boy, he was known as Joshua Ben Joseph. As a young man, he was known as Jesus of Nazareth. As he opened his heart and mind and allowed the wisdom and love of God to be expressed in him and through him, he became the Christ, which is a title for the unconditional love of God. As he fully became aware of his mission, he freely joined his will with the will of God the Father. He defeated Lucifer, Satan and the (prince of this world, whom we now know as the devil).

For about 2,000 years, the world has been given time to ingest the loving teachings of Christ. The world is on the threshold of entering into the stage called the light. Therein humanity will become more spiritually enlightened, and as the quarantine, that we have been under since the rebellion in heaven is removed. Humanity will have to make many adjustments. We can do it the easy way, or we can do it the hard way. Unfortunately, it appears that humanity is still a stiff-necked people hanging onto our self-glorification, self-aggrandizement, and self-gratification. Humanity will go through the trials and tribulations of the spiritual, mental, and physical attitude adjustment needed to be able to enter that light. The time for that light to arrive is nigh at hand, and it is human kind's decision to enter the light or not.

We are all spiritual sons and daughters of God invested with the authority over this world. When all humanity wants and seeks world peace, it can be found, but only when all humanity seeks it. Therefore, let us all seek with all our hearts and minds for that peace. It can and will be found.

God planned it that way.

Made in the USA
Monee, IL
04 November 2026

67384280R00108